Prairie Creek Public Library District
501 Carriage House Lane
Dwight, IL 60420
Call to renew: (815) 584-3061

SCIENTIFIC AMERICAN™

Critical Anthologies on Environment and Climate™

CRITICAL PERSPECTIVES ON
ENERGY AND POWER

Edited by Linley Erin Hall

The Rosen Publishing Group, Inc., New York

Published in 2007 by The Rosen Publishing Group, Inc.
29 East 21st Street, New York, NY 10010

The articles in this book first appeared in the pages of *Scientific American*, as follows: "The End of Cheap Oil" by Colin J. Campbell and Jean H. Laherrère, March 1998; "Oil Production in the 21st Century" by Roger N. Anderson, March 1998; "Liquid Fuels from Natural Gas" by Safaa A. Fouda, March 1998; "Next-Generation Nuclear Power" by James A. Lake, Ralph G. Bennett, and John F. Kotek, January 2002; "Disposing of Nuclear Waste" by the Editors, September 1995; "Burial of Radioactive Waste Under the Seabed" by Charles D. Hollister and Steven Nadis, January 1998; "Fusion" by Harold P. Furth, September 1995; "On the Road to Fuel-Cell Cars" by Steven Ashley, March 2005; "The Power Plant in Your Basement" by Alan C. Lloyd, July 1999; "Power to the People" by David Schneider, May 1997; "Flammable Ice" by Erwin Suess, Gerhard Bohrmann, Jens Greinert, and Erwin Lausch, November 1999; "Solar Energy" by William Hoagland, September 1995; "Thermophotovoltaics" by Timothy J. Coutts and Mark C. Fitzgerald, September 1998; "Change in the Wind" by W. Wayt Gibbs, October 1997.

First Edition

Library of Congress Cataloging-in-Publication Data

Critical perspectives on energy and power/edited by Linley Erin Hall.
    p. cm.—(Scientific American critical anthologies on environment and climate)
Includes bibliographical references and index.
ISBN 1-4042-0689-2 (library binding)
1. Power resources. I. Hall, Linley Erin. II. Series.

TJ163.2C77 2006
333.79—dc22

                                                2005033237

Manufactured in the United States of America

On the cover: Building engineer Ernst Loubeau performs maintenance on a solar power cell mounted on the roof of the Field Museum of Natural History in Chicago, Illinois.

CONTENTS

Introduction

The Energy Information Administration, an independent agency of the U.S. Department of Energy, predicts that the global energy demand will increase 57 percent between 2002 and 2025. Much of this increase will be due to developing nations, whose energy needs are expected to double. The rising demand for energy takes into account growing populations as well as changes in developing nations, such as electrification of rural areas and increased manufacturing.

While energy produced from both fossil fuels and other sources will grow, fossil fuel use will increase more rapidly than production, according to the Energy Information Administration. In particular, natural gas consumption is estimated to increase by 69 percent. Oil and coal consumption will both increase as well.

Reliable sources of energy are crucial to maintaining accepted standards of living in industrialized nations. Fossil fuels are a finite resource, however. One day they will be gone, and before that day arrives, they will become increasingly more costly to extract from the earth. As oil, gas, and coal become more scarce, their selling price will also increase, putting an enormous strain on the budgets of both businesses and average citizens. Fossil fuels also take a toll on the environment. Drilling for oil and mining coal can be harmful to the land, water, and ecosystems surrounding these operations. In addition, the burning of fossil fuels releases greenhouse gases and other pollutants into the atmosphere, contributing to air pollution and global warming.

Energy alternatives are many, but, like fossil fuels, each one has its drawbacks. Nuclear power does not pollute the air, but it does generate large quantities of radioactive waste that for many years remains dangerous—and potentially fatal—to living organisms, including humans, that come into contact with it. Many people also fear accidents at nuclear plants that release toxic radiation. For this reason, few people want nuclear power plants built and operated near their communities. New technologies may be able to allay such fears. In addition, if researchers can make nuclear fusion work on a large scale, these plants would be safer in many ways than the current fission reactors.

Fuel cells, which convert a fuel such as hydrogen to electricity, are leading candidates to provide power to a wide variety of mobile and remote technologies in the near future. The world's automakers are developing vehicles that run on fuel cells rather than gasoline, and one can soon expect to see them powering laptops, personal digital assistants, and other electronic devices. Large fuel cells can even power homes and other buildings. Since the hydrogen that powers fuel cells is currently made from fossil fuels, however, they are not an entirely green, or environmentally friendly, technology.

Green technologies include solar, wind, and hydroelectric power. More energy from the sun falls on the earth each day than the more than 6 billion people on the planet could use in twenty-seven years, according to the National Renewable Energy Laboratory. Humans cannot tap into all of that available energy because much of it is used by plants for photosynthesis or falls on out-of-the-way areas, but solar power nevertheless has great potential as an energy source. Wind energy is also gaining in popularity as windmill farms continue to be erected in some of the most consistently windy areas of the United States.

Most hydroelectric power plants use a dam on a river to store water in a reservoir. When water is released from the reservoir, its rushing force spins turbines, thus generating electricity. Dams have negative ecological repercussions, however, as they flood land upstream of the dam to create the reservoir and inhibit the flow of the river downstream. This alteration of water flow patterns changes ecosystems, often causing species to die out or migrate elsewhere. As such, a movement is under way to remove already existing dams in many locations.

As always, economics is an important factor in any consideration of energy production and consumption. Energy derived from alternative and renewable sources is generally more expensive than that produced from traditional fossil fuel sources. Some environmentally aware customers in the developed world are willing to pay higher prices for clean alternative energy, but, thus far, most consumers have not proven as willing. To continue their still-fragile process of economic and industrial development, third world nations need cheap energy. Further research into and development of renewable energy is needed to bring down prices, although rising oil costs are now narrowing the gap between the price of fossil and alternative fuels.

As the world waits for clean energy technologies to become more efficient and affordable, technologies that are less polluting are becoming more prevalent. Hybrid vehicles, for example, combine an electric motor with a gas-powered internal combustion engine, using each at different times. As such, the hybrid vehicles get much better gas mileage. In addition to hybrid-only vehicles such as the Toyota Prius, several auto manufacturers now offer hybrid versions of some of their conventional vehicles. These include luxury cars, pickup trucks, and SUVs.

Many potentially useful technologies are not being pursued, however, because the United States and other nations have decided to focus their energy research elsewhere. Disposal of radioactive waste under the seabed, thermophotovoltaics, and other topics covered in this volume have not progressed significantly since the articles were written because of lack of funding.

Obviously, a balance needs to be struck between meeting the world's increasing energy needs and protecting the world's increasingly threatened and fragile environment. Different individuals, politicians, businesses, industries, and nations have wildly varying ideas of what this balance should look like. Unfortunately, many people—including politicians who have a role in energy policymaking—do not have the scientific or economic background needed to understand the issues that arise when energy and environment intersect. Education on these issues will be an important first step to truly finding the right balance between the human need for energy and the fragility of the environment that is our home and sustenance. —*LEH*

1 Fossil Fuels

Americans have become used to having what seems like an endless and, until recently, relatively inexpensive supply of oil to produce gasoline, power electrical plants, and serve as a raw material for plastics and other consumer products. The average petroleum demand in the United States in 2004 was more than 20 million barrels per day, more than half of which was imported, according to the U.S. Department of Energy.

The days of cheap, plentiful oil may soon be at an end, however. In the last thirty years, consumers have endured gasoline shortages and price hikes due to political tensions, like the Arab oil embargoes in the 1970s, and natural disasters, such as 2005's Hurricanes Katrina and Rita, which damaged oil refineries along the Gulf Coast and disrupted supplies. The scarcities and price hikes associated with these situations were temporary; those resulting from the coming oil shortage will not be.

Oil, like all fossil fuels, is a finite resource. Only so much is accessible, and only part of that can be obtained easily and cheaply. Researchers *monitoring the oil industry suggest that all the easily accessible oil has already been found. Soon it will become more difficult, and thus more expensive, to extract the remaining oil from the earth. The following article explains why this will happen, how soon it will happen, and what the potential consequences are. —LEH*

"The End of Cheap Oil"
by Colin J. Campbell and Jean H. Laherrère
Scientific American, March 1998

In 1973 and 1979 a pair of sudden price increases rudely awakened the industrial world to its dependence on cheap crude oil. Prices first tripled in response to an Arab embargo and then nearly doubled again when Iran dethroned its Shah, sending the major economies sputtering into recession. Many analysts warned that these crises proved that the world would soon run out of oil. Yet they were wrong.

Their dire predictions were emotional and political reactions; even at the time, oil experts knew that they had no scientific basis. Just a few years earlier oil explorers had discovered enormous new oil provinces on the north slope of Alaska and below the North Sea off the coast of Europe. By 1973 the world had consumed, according to many experts' best estimates, only about one eighth of its endowment of readily accessible crude oil (so-called conventional oil). The five Middle Eastern members of the Organization of

Petroleum Exporting Countries (OPEC) were able to hike prices not because oil was growing scarce but because they had managed to corner 36 percent of the market. Later, when demand sagged, and the flow of fresh Alaskan and North Sea oil weakened OPEC's economic stranglehold, prices collapsed.

The next oil crunch will not be so temporary. Our analysis of the discovery and production of oil fields around the world suggests that within the next decade, the supply of conventional oil will be unable to keep up with demand. This conclusion contradicts the picture one gets from oil industry reports, which boasted of 1,020 billion barrels of oil (Gbo) in "proved" reserves at the start of 1998. Dividing that figure by the current production rate of about 23.6 Gbo a year might suggest that crude oil could remain plentiful and cheap for 43 more years—probably longer, because official charts show reserves growing.

Unfortunately, this appraisal makes three critical errors. First, it relies on distorted estimates of reserves. A second mistake is to pretend that production will remain constant. Third and most important, conventional wisdom erroneously assumes that the last bucket of oil can be pumped from the ground just as quickly as the barrels of oil gushing from wells today. In fact, the rate at which any well—or any country—can produce oil always rises to a maximum and then, when about half the oil is gone, begins falling gradually back to zero.

From an economic perspective, when the world runs completely out of oil is thus not directly relevant:

what matters is when production begins to taper off. Beyond that point, prices will rise unless demand declines commensurately. Using several different techniques to estimate the current reserves of conventional oil and the amount still left to be discovered, we conclude that the decline will begin before 2010.

Digging for the True Numbers

We have spent most of our careers exploring for oil, studying reserve figures and estimating the amount of oil left to discover, first while employed at major oil companies and later as independent consultants. Over the years, we have come to appreciate that the relevant statistics are far more complicated than they first appear.

Consider, for example, three vital numbers needed to project future oil production. The first is the tally of how much oil has been extracted to date, a figure known as cumulative production. The second is an estimate of reserves, the amount that companies can pump out of known oil fields before having to abandon them. Finally, one must have an educated guess at the quantity of conventional oil that remains to be discovered and exploited. Together they add up to ultimate recovery, the total number of barrels that will have been extracted when production ceases many decades from now.

The obvious way to gather these numbers is to look them up in any of several publications. That approach works well enough for cumulative production statistics because companies meter the oil as it flows from their wells. The record of production is not perfect (for example, the two billion barrels of Kuwaiti oil wastefully burned by Iraq in 1991 is usually not included in official statistics), but errors are relatively easy to spot and rectify. Most experts agree that the industry had removed just over 800 Gbo from the earth at the end of 1997.

Getting good estimates of reserves is much harder, however. Almost all the publicly available statistics are taken from surveys conducted by the *Oil and Gas Journal* and *World Oil*. Each year these two trade journals query oil firms and governments around the world. They then publish whatever production and reserve numbers they receive but are not able to verify them.

The results, which are often accepted uncritically, contain systematic errors. For one, many of the reported figures are unrealistic. Estimating reserves is an inexact science to begin with, so petroleum engineers assign a probability to their assessments. For example, if, as geologists estimate, there is a 90 percent chance that the Oseberg field in Norway contains 700 million barrels of recoverable oil but only a 10 percent chance that it will yield 2,500 million more barrels, then the lower figure should be cited as the so-called P90 estimate (P90 for "probability 90 percent") and the higher as the P10 reserves.

In practice, companies and countries are often deliberately vague about the likelihood of the reserves they report, preferring instead to publicize whichever figure, within a P10 to P90 range, best suits them. Exaggerated estimates can, for instance, raise the price of an oil company's stock.

The members of OPEC have faced an even greater temptation to inflate their reports because the higher their reserves, the more oil they are allowed to export. National companies, which have exclusive oil rights in the main OPEC countries, need not (and do not) release detailed statistics on each field that could be used to verify the country's total reserves. There is thus good reason to suspect that when, during the late 1980s, six of the 11 OPEC nations increased their reserve figures by colossal amounts, ranging from 42 to 197 percent, they did so only to boost their export quotas.

Previous OPEC estimates, inherited from private companies before governments took them over, had probably been conservative, P90 numbers. So some upward revision was warranted. But no major new discoveries or technological breakthroughs justified the addition of a staggering 287 Gbo. That increase is more than all the oil ever discovered in the U.S.— plus 40 percent. Non-OPEC countries, of course, are not above fudging their numbers either: 59 nations stated in 1997 that their reserves were unchanged from 1996. Because reserves naturally drop as old fields are drained and jump when new fields are discovered, perfectly stable numbers year after year are implausible.

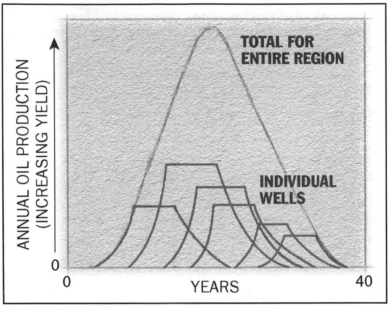

Flow of oil starts to fall from any large region when about half the crude is gone. Adding the output of fields of various sizes and ages usually yields a bell-shaped production curve for the region as a whole. M. King Hubbert, a geologist with Shell Oil, exploited this fact in 1956 to predict correctly that oil from the lower 48 American states would peak around 1969.

Unproved Reserves

Another source of systematic error in the commonly accepted statistics is that the definition of reserves varies widely from region to region. In the U.S., the Securities and Exchange Commission allows companies to call reserves "proved" only if the oil lies near a producing well and there is "reasonable certainty" that it can be recovered profitably at current oil prices, using existing technology. So a proved reserve estimate in the U.S. is roughly equal to a P90 estimate.

Regulators in most other countries do not enforce particular oil-reserve definitions. For many years, the former Soviet countries have routinely released wildly optimistic figures—essentially P10 reserves. Yet analysts have often misinterpreted these as estimates of "proved" reserves. *World Oil* reckoned reserves in the former Soviet Union amounted to 190 Gbo in 1996, whereas the *Oil and Gas Journal* put the number at 57 Gbo. This large discrepancy shows just how elastic these numbers can be.

Using only P90 estimates is not the answer, because adding what is 90 percent likely for each field, as is done in the U.S., does not in fact yield what is 90 percent likely for a country or the entire planet. On the contrary, summing many P90 reserve estimates always understates the amount of proved oil in a region. The only correct way to total up reserve numbers is to add the mean, or average, estimates of oil in each field. In practice, the median estimate, often called "proved and probable," or P50 reserves, is more widely used and is good enough. The P50 value is the number of barrels of oil that are as likely as not to come out of a well during its lifetime, assuming prices remain within a limited range. Errors in P50 estimates tend to cancel one another out.

We were able to work around many of the problems plaguing estimates of conventional reserves by using a large body of statistics maintained by Petroconsultants in Geneva. This information, assembled over 40 years from myriad sources, covers some 18,000 oil fields worldwide. It, too, contains some dubious reports, but we did our best to correct these sporadic errors.

According to our calculations, the world had at the end of 1996 approximately 850 Gbo of conventional oil in P50 reserves—substantially less than the 1,019 Gbo reported in the *Oil and Gas Journal* and the 1,160 Gbo estimated by *World Oil*. The difference is actually greater than it appears because our value represents the amount most likely to come out of known oil fields, whereas the larger number is supposedly a cautious estimate of proved reserves.

For the purposes of calculating when oil production will crest, even more critical than the size of the world's reserves is the size of ultimate recovery—all the cheap oil there is to be had. In order to estimate that, we need to know whether, and how fast, reserves are moving up or down. It is here that the official statistics become dangerously misleading.

Diminishing Returns

According to most accounts, world oil reserves have marched steadily upward over the past 20 years. Extending that apparent trend into the future, one could easily conclude, as the U.S. Energy Information Administration has, that oil production will continue to rise unhindered for decades to come, increasing almost two thirds by 2020.

Such growth is an illusion. About 80 percent of the oil produced today flows from fields that were found before 1973, and the great majority of them are declining. In the 1990s oil companies have discovered an average of seven Gbo a year; last year they drained more than three times as much. Yet official figures indicated that proved reserves did not fall by 16 Gbo, as one would expect—rather they *expanded* by 11 Gbo. One reason is that several dozen governments opted not to report declines in their reserves, perhaps to enhance their political cachet and their ability to obtain loans. A more important cause of the expansion lies in revisions: oil companies replaced earlier estimates of the reserves left in many fields with higher numbers. For most purposes, such amendments are harmless, but they seriously distort forecasts extrapolated from published reports.

To judge accurately how much oil explorers will uncover in the future, one has to backdate every revision to the year in which the field was first discovered— not to the year in which a company or country corrected an earlier estimate. Doing so reveals that global discovery peaked in the early 1960s and has been falling steadily ever since. By extending the trend to zero, we can make a good guess at how much oil the industry will ultimately find.

We have used other methods to estimate the ulti- mate recovery of conventional oil for each country [*see boxes on pages 22–25*], and we calculate that the oil industry will be able to recover only about another

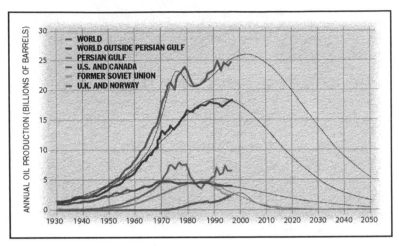

Global production of oil, both conventional and unconventional, recovered after falling in 1973 and 1979. But a more permanent decline is less than 10 years away, according to the authors' model, based in part on multiple Hubbert curves (*lighter lines*). U.S. and Canadian oil topped out in 1972; production in the former Soviet Union has fallen 45 percent since 1987. A crest in the oil produced outside the Persian Gulf region now appears imminent.

1,000 billion barrels of conventional oil. This number, though great, is little more than the 800 billion barrels that have already been extracted.

It is important to realize that spending more money on oil exploration will not change this situation. After the price of crude hit all-time highs in the early 1980s, explorers developed new technology for finding and recovering oil, and they scoured the world for new fields. They found few: the discovery rate continued its decline uninterrupted. There is only so much crude oil in the world, and the industry has found about 90 percent of it.

Predicting the Inevitable

Predicting when oil production will stop rising is
relatively straightforward once one has a good estimate
of how much oil there is left to produce. We simply
apply a refinement of a technique first published in
1956 by M. King Hubbert. Hubbert observed that in any
large region, unrestrained extraction of a finite resource
rises along a bell-shaped curve that peaks when about
half the resource is gone. To demonstrate his theory,
Hubbert fitted a bell curve to production statistics and
projected that crude oil production in the lower 48
U.S. states would rise for 13 more years, then crest in
1969, give or take a year. He was right: production
peaked in 1970 and has continued to follow Hubbert
curves with only minor deviations. The flow of oil from
several other regions, such as the former Soviet Union
and the collection of all oil producers outside the Middle
East, also follows Hubbert curves quite faithfully.

The global picture is more complicated, because
the Middle East members of OPEC deliberately reined
back their oil exports in the 1970s, while other nations
continued producing at full capacity. Our analysis
reveals that a number of the largest producers, including
Norway and the U.K., will reach their peaks around
the turn of the millennium unless they sharply curtail
production. By 2002 or so the world will rely on Middle
East nations, particularly five near the Persian Gulf
(Iran, Iraq, Kuwait, Saudi Arabia and the United Arab
Emirates), to fill in the gap between dwindling supply

and growing demand. But once approximately 900 Gbo have been consumed, production must soon begin to fall. Barring a global recession, it seems most likely that world production of conventional oil will peak during the first decade of the 21st century.

Perhaps surprisingly, that prediction does not shift much even if our estimates are a few hundred billion

How Much Oil Is Left to Find?

We combined several techniques to conclude that about 1,000 billion barrels of conventional oil remain to be produced. First, we extrapolated published production figures for older oil fields that have begun to decline. The Thistle field off the coast of Britain, for example, will yield about 420 million barrels (a). Second, we plotted the

We can predict the amount of remaining oil from the decline of aging fields . . .

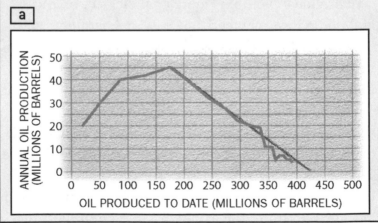

ANNUAL OIL PRODUCTION (MILLIONS OF BARRELS)

OIL PRODUCED TO DATE (MILLIONS OF BARRELS)

barrels high or low. Craig Bond Hatfield of the University of Toledo, for example, has conducted his own analysis based on a 1991 estimate by the U.S. Geological Survey of 1,550 Gbo remaining—55 percent higher than our figure. Yet he similarly concludes that the world will hit maximum oil production within the next 15 years. John D. Edwards of the University of Colorado published

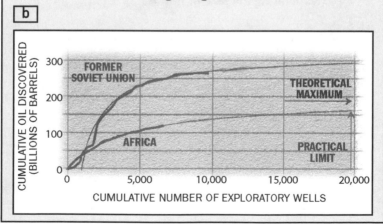

amount of oil discovered so far in some regions against the cumulative number of exploratory wells drilled there. Because larger fields tend to be found first—they are simply too large to miss—the curve rises rapidly and then flattens, eventually reaching a theoretical maximum: for Africa, 192 Gbo. But the time and cost of exploration impose a more practical limit of perhaps 165 Gbo (*b*).

... from the diminishing returns on exploration in larger regions ...

b

CUMULATIVE OIL DISCOVERED (BILLIONS OF BARRELS)

FORMER SOVIET UNION

THEORETICAL MAXIMUM

AFRICA

PRACTICAL LIMIT

300

200

100

0

0   5,000   10,000   15,000   20,000

CUMULATIVE NUMBER OF EXPLORATORY WELLS

continued on following page

last August one of the most optimistic recent estimates of oil remaining: 2,036 Gbo. (Edwards concedes that the industry has only a 5 percent chance of attaining that very high goal.) Even so, his calculations suggest that conventional oil will top out in 2020.

Smoothing the Peak

Factors other than major economic changes could speed or delay the point at which oil production begins to

continued from previous page

Third, we analyzed the distribution of oil-field sizes in the Gulf of Mexico and other provinces. Ranked according to size and then graphed on a logarithmic scale, the fields tend to fall along a parabola that grows predictably over time (c). (Interestingly, galaxies, urban populations and other natural agglomerations also seem to fall along

... by extrapolating the size of new fields into the future ...

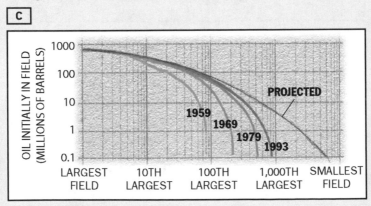

decline. Three in particular have often led economists and academic geologists to dismiss concerns about future oil production with naive optimism.

First, some argue, huge deposits of oil may lie undetected in far-off corners of the globe. In fact, that is very unlikely. Exploration has pushed the frontiers back so far that only extremely deep water and polar regions remain to be fully tested, and even their prospects are now reasonably well understood.

such parabolas.) Finally, we checked our estimates by matching our projections for oil production in large areas, such as the world outside the Persian Gulf region, to the rise and fall of oil discovery in those places decades earlier (d).

—C. J. C. and J. H. L.

... and by matching production to earlier discovery trends.

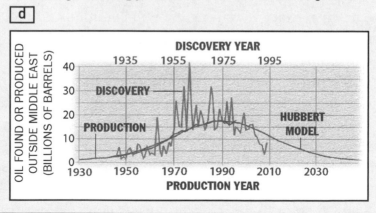

d

Theoretical advances in geochemistry and geophysics have made it possible to map productive and prospective fields with impressive accuracy. As a result, large tracts can be condemned as barren. Much of the deepwater realm, for example, has been shown to be absolutely nonprospective for geologic reasons.

What about the much touted Caspian Sea deposits? Our models project that oil production from that region will grow until around 2010. We agree with analysts at the USGS World Oil Assessment program and elsewhere who rank the total resources there as roughly equivalent to those of the North Sea—that is, perhaps 50 Gbo but certainly not several hundreds of billions as sometimes reported in the media.

A second common rejoinder is that new technologies have steadily increased the fraction of oil that can be recovered from fields in a basin—the so-called recovery factor. In the 1960s oil companies assumed as a rule of thumb that only 30 percent of the oil in a field was typically recoverable; now they bank on an average of 40 or 50 percent. That progress will continue and will extend global reserves for many years to come, the argument runs.

Of course, advanced technologies will buy a bit more time before production starts to fall [see "Oil Production in the 21st Century," by Roger N. Anderson, on page 32]. But most of the apparent improvement in recovery factors is an artifact of reporting. As oil fields grow old, their owners often deploy newer technology to slow their decline. The falloff also allows engineers

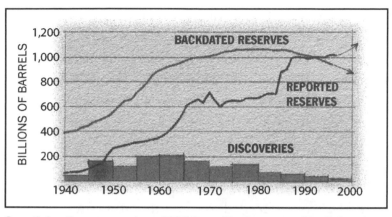

Growth in oil reserves since 1980 is an illusion caused by belated corrections to oil-field estimates. Backdating the revisions to the year in which the fields were discovered reveals that reserves have been falling because of a steady decline in newfound oil (*lighter gray*).

to gauge the size of the field more accurately and to correct previous underestimation—in particular P90 estimates that by definition were 90 percent likely to be exceeded.

Another reason not to pin too much hope on better recovery is that oil companies routinely count on technological progress when they compute their reserve estimates. In truth, advanced technologies can offer little help in draining the largest basins of oil, those onshore in the Middle East where the oil needs no assistance to gush from the ground.

Last, economists like to point out that the world contains enormous caches of unconventional oil that can substitute for crude oil as soon as the price rises high enough to make them profitable. There is no question that the resources are ample: the Orinoco oil belt in Venezuela has been assessed to contain a staggering 1.2 trillion barrels of the sludge known as heavy oil. Tar sands and shale deposits in Canada and the former Soviet Union may contain the equivalent of more than 300 billion barrels of oil. Theoretically, these unconventional oil reserves could quench the world's thirst for liquid fuels as conventional oil passes its prime. But the industry will be hard-pressed for the time and money needed to ramp up production of unconventional oil quickly enough.

Such substitutes for crude oil might also exact a high environmental price. Tar sands typically emerge from strip mines. Extracting oil from these sands and shales creates air pollution. The Orinoco sludge contains heavy metals and sulfur that must be removed. So governments may restrict these industries from growing as fast as they could. In view of these potential obstacles, our skeptical estimate is that only 700 Gbo will be produced from unconventional reserves over the next 60 years.

On the Down Side

Meanwhile global demand for oil is currently rising at more than 2 percent a year. Since 1985, energy use is up about 30 percent in Latin America, 40 percent in Africa and 50 percent in Asia. The Energy Information Administration forecasts that worldwide demand for oil will increase 60 percent (to about 40 Gbo a year) by 2020.

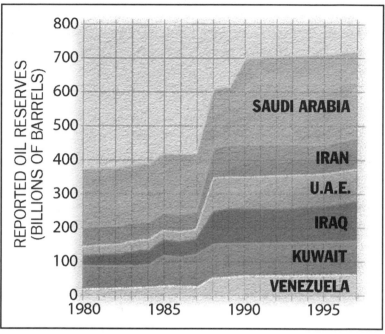

Suspicious jump in reserves reported by six OPEC members added 300 billion barrels of oil to official reserve tallies yet followed no major discovery of new fields.

The switch from growth to decline in oil production will thus almost certainly create economic and political tension. Unless alternatives to crude oil quickly prove themselves, the market share of the OPEC states in the Middle East will rise rapidly. Within two years, these nations' share of the global oil business will pass 30 percent, nearing the level reached during the oil-price shocks of the 1970s. By 2010 their share will quite probably hit 50 percent.

The world could thus see radical increases in oil prices. That alone might be sufficient to curb demand, flattening production for perhaps 10 years. (Demand fell more than 10 percent after the 1979 shock and took 17 years to recover.) But by 2010 or so, many Middle Eastern nations will themselves be past the midpoint. World production will then have to fall.

With sufficient preparation, however, the transition to the post-oil economy need not be traumatic. If advanced methods of producing liquid fuels from natural gas can be made profitable and scaled up quickly, gas could become the next source of transportation fuel [see "Liquid Fuels from Natural Gas," by Safaa A. Fouda, on page 47]. Safer nuclear power, cheaper renewable energy, and oil conservation programs could all help postpone the inevitable decline of conventional oil.

Countries should begin planning and investing now. In November a panel of energy experts appointed by President Bill Clinton strongly urged the administration to increase funding for energy research by $1 billion over the next five years. That is a small step in the right direction, one that must be followed by giant leaps from the private sector.

The world is not running out of oil—at least not yet. What our society does face, and soon, is the end of the abundant and cheap oil on which all industrial nations depend.

The Authors

Colin J. Campbell and Jean H. Laherrère have each worked in the oil industry for more than 40 years. After completing his Ph.D. in geology at the University of Oxford, Campbell worked for Texaco as an exploration geologist and then at Amoco as chief geologist for Ecuador. His decade long study of global oil-production trends has led to two books and numerous papers. Laherrère's early work on seismic refraction surveys contributed to the discovery of Africa's largest oil field. At Total, a French oil company, he supervised exploration techniques worldwide. Both Campbell and Laherrère are currently associated with Petroconsultants in Geneva.

The previous article paints a grim picture of the future of oil production and consumption, but not everyone is dismayed by the prospects for the oil industry. Researchers are developing a wide variety of new technologies to help improve oil recovery, some of which are discussed in the following article.

One of these techniques is the injection of carbon dioxide, steam, or natural gas into oil wells. The gas is supposed to displace oil in the underground rock formations, pushing it to the surface. Injecting carbon dioxide into oil fields is potentially good for the environment, since it sequesters the greenhouse gas underground, out of the atmosphere. Because of the expense, however, oil companies are more likely to use *cheaply available water steam rather than carbon dioxide.*

The other techniques described in the following article—4-D seismic surveying, directional drilling, and tapping oil fields deep underwater—have been implemented to varying degrees by different companies, yet it is still too early to determine what their long-term impact on the oil industry, if any, will be. —LEH

"Oil Production in the 21st Century"
by Roger N. Anderson
Scientific American, March 1998

On the face of it, the outlook for conventional oil—the cheap, easily recovered crude that has furnished more than 95 percent of all oil to date—seems grim. In 2010, according to forecasts, the world's oil-thirsty economies will demand about 10 billion more barrels than the industry will be able to produce. A supply shortfall that large, equal to almost half of all the oil extracted in 1997, could lead to price shocks, economic recession and even wars.

Fortunately, four major technological advances are ready to fill much of the gap by accelerating the discovery of new oil reservoirs and by dramatically increasing the fraction of oil within existing fields that can be removed economically, a ratio known as the recovery factor. These technologies could lift global oil production rates more than 20 percent by 2010 if they are deployed as planned on the largest oil fields

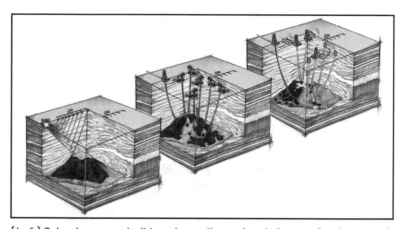

(*Left*) Seismic survey builds a three-dimensional picture of underground strata one vertical slice at a time. Sound waves generated at the surface ricochet off boundaries between layers of ordinary rock and those bearing oil, water or gas. The returning sounds are picked up by a string of microphones. Computers later translate the patterns into images and ultimately into a model that guides the drilling of wells. (*Center*) Production wells often draw water from below and gas from above into pore spaces once full of oil. This complex flow strands pockets of crude far from wells; traditional drilling techniques thus miss up to two thirds of the oil in a reservoir. But repeated seismic surveys can now be assembled into a 4-D model that not only tracks where oil, gas and water in the field are located but also predicts where they will go next. Advanced seismic monitoring works well on about half the world's oil fields, but it fails on oil buried in very hard rock or beneath beds of salt (*thick white layer*). (*Right*) Injection of liquid carbon dioxide can rejuvenate dying oil fields. Pumped at high pressure from tanks into wells that have ceased producing oil, the carbon dioxide flows through the reservoir and, if all goes well, pushes the remaining oil down toward active wells. Steam and natural gas are sometimes also used for this purpose. Alternatively, water can be injected below a pocket of by-passed crude in order to shepherd the oil into a well. In the future, "smart" wells currently under development will be able to retrieve oil simultaneously from some branches of the well while using other branches to pump water out of the oil stream and back into the formation from which it came.

within three to five years. Such rapid adoption may seem ambitious for an industry that traditionally has taken 10 to 20 years to put new inventions to use. But in this case, change will be spurred by formidable economic forces.

For example, in the past two years, the French oil company Elf has discovered giant deposits off the coast of West Africa. In the same period, the company's stock doubled, as industry analysts forecasted that Elf's production would increase by 8 percent in 2001. If the other major oil producers follow suit, they should be able by 2010 to provide an extra five billion barrels of oil each year, closing perhaps half the gap between global supply and demand.

This article will cover the four advances in turn, beginning with a new way to track subterranean oil.

Tracking Oil in Four Dimensions

Finding oil became much more efficient after 1927, when geologists first successfully translated acoustic reflections into detailed cross sections of the earth's crust. Seismologists later learned how to piece together several such snapshots to create three-dimensional models of the oil locked inside layers of porous rock. Although this technique, known as 3-D seismic analysis, took more than a decade to become standard practice, it is now credited with increasing oil discovery and recovery rates by 20 percent.

In recent years, scientists in my laboratory at Columbia University and elsewhere have developed

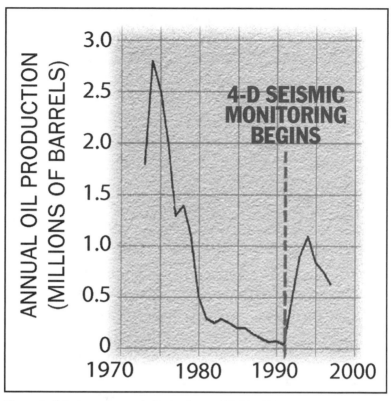

Flow of oil from a reservoir in the largest field off the Louisiana shore resurged in 1992, shortly after operators began using 4-D seismic monitoring to locate hidden caches of oil.

even more powerful techniques capable of tracking the movement of oil, gas and water as drilled wells drain the subterranean strata—a "4-D" scheme that includes the added dimension of time. This information can then be used to do a "what if" analysis on the oil field, designing ways to extract as much of the oil as quickly and cheaply as possible.

Compared with its predecessor, the 4-D approach seems to be catching on quickly: the number of oil fields benefiting from it has doubled in each of the past four years and now stands at about 60. Such monitoring can boost recovery factors by 10 to 15 percentage points. Unfortunately, the technique will work in only about half the world's major fields, those where relatively soft rock is suffused with oil and natural gas.

Gassing Things Up

When geologists began studying the new time-lapse measurements, they were surprised to discover that one of the most basic notions about oil movement—that it naturally settles between lighter gas above and heavier groundwater below—oversimplifies the behavior of real oil fields. In fact, most wells produce complex, fractal drainage patterns that cause the oil to mix with gas and water. As a result, specialists now know that the traditional technique of pumping a well until the oil slows to a trickle often leaves 60 percent or more of the oil behind.

A more efficient strategy is to pump natural gas, steam or liquid carbon dioxide into dead wells. The infusion then spreads downward through pores in the rock and, if one has planned carefully, pushes oil that otherwise would have been abandoned toward a neighboring well. Alternatively, water is often pumped below the oil to increase its pressure, helping it flow up to the surface.

Injections of steam and carbon dioxide have been shown to increase recovery factors by 10 to 15 percentage points. Unfortunately, they also raise the cost of oil production by 50 to 100 percent—and that added expense falls on top of a 10 to 25 percent surcharge for 4-D seismic monitoring. So unless carbon dioxide becomes much cheaper (perhaps because global-warming treaties restrict its release) these techniques will probably continue to serve only as a last resort.

Steering to Missed Oil

A third major technological advance, known as directional drilling can tap bypassed deposits of oil at less

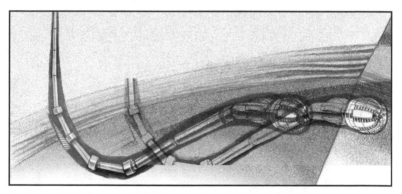

Horizontal drilling was impractical when oil rigs had to rotate the entire drill string—up to 5,800 meters (roughly 19,000 feet) of it—in order to turn the rock-cutting bit at the bottom. Wells that swing 90 degrees over a space of just 100 meters are now common thanks to the development of motors that can run deep underground. The motor's driveshaft connects to the bit through a transmission in a bent section of pipe. The amount of bend determines how tight a curve the drill will carve; drillers can twist the string to control the direction of the turn.

Drilling console allows an engineer at the surface to monitor sensors near the drill bit that indicate whether it has hit oil or water. The drill can then be steered into position for the optimum yield.

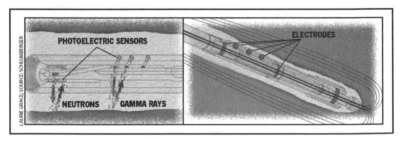

Sensors near the bit can detect oil, water and gas. One device measures the porosity of the surrounding rock by emitting neutrons, which scatter off hydrogen atoms. Another takes a density reading by shooting out gamma rays that interact with adjacent electrons. Oil and water also affect electrical resistance measured from a current passed through the bit, the rock and nearby electrodes.

expense than injection. Petroleum engineers can use a variety of new equipment to swing a well from vertical to entirely horizontal within a reservoir several kilometers underground.

Traditionally, drillers rotated the long steel pipe, or "string," that connects the rig at the surface to the bit at the bottom of the well. That method fails when the pipe must turn a corner—the bend would break the

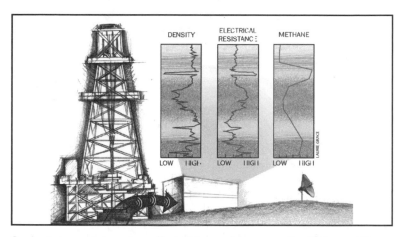

Geologic measurements collected by sensors near the bottom of the drill pipe can be analyzed at the wellhead or transmitted via satellite to engineers anywhere in the world. Several characteristics of the rocks surrounding the drill bit can reveal the presence of oil or gas (*above*). Petroleum tends to accumulate in relatively light, porous rocks, for example, so some geosteering systems calculate the bulk density of nearby strata. Others measure the electrical resistance of the earth around the drill; layers soaked with briny water have a much lower resistance than those rich in oil. Gas chromatographs at the surface analyze the returning flow of lubricating mud for natural gas captured during its journey.

Advanced drills use mud pumped through the inside of the string to rotate the hit, to communicate sensor measurements and to carry rock fragments out of the well. On its way down, the mud first enters a rotating valve (*a*) which converts data radioed to the tool from various sensors into surges in the mud stream. (At the surface, the pulses are translated back into a digital signal of up to 10 bits per second.) The mud next flows into a motor. A spiral driveshaft fits inside the helical motor casing in a way that creates chambers (*b*). As the cavities fill with mud, the shaft turns in order to relieve the hydraulic pressure. The mud finally exits through the rotating bit and returns to the surface, with fresh cuttings cleared from near the bit.

rotating string. So steerable drill strings do not rotate; instead a mud-driven motor inserted near the bit turns only the diamond-tipped teeth that do the digging. An elbow of pipe placed between the mud motor and the bit controls the direction of drilling.

Threading a hole through kilometers of rock into a typical oil zone 30 meters (about 100 feet) thick is precise work. Schlumberger, Halliburton and other international companies have developed sophisticated sensors that significantly improve the accuracy of

"Smart" wells of the near future (*a*) will use computers and water monitors near the bottom of the well to detect dilution of the oil stream by water. Hydrocyclonic separators will then shunt the water into a separate branch of the well that empties beneath the oil reservoir. Forked wells (*b*) can extract oil from several oil-bearing layers at once. Computer-controlled chokes inserted in the well pipe maintain the optimum flow of oil to the surface.

drilling. These devices which operates at depths of up to 6,000 meters and at temperatures as high as 200 degrees Celsius (400 degrees Fahrenheit), attach to the drill pipe just above or below the mud motor. Some measure the electrical resistance of the surrounding rock. Others send out neutrons and gamma rays; then they count the number that are scattered back by the rock and pore fluids. These measurements and the current position of the bit (calculated by an inertial guidance system) are sent back to the surface through pulses in the flow of the very mud used to turn the motor and lubricate the well bore. Engineers can adjust the path of the drill accordingly, thus snaking their way to the most oil-rich part of the formation.

Once the hole is completed, drillers typically erect production equipment on top of the wellhead. But several companies are now developing sensors that can detect the mix of oil, gas and water near its point of entry deep within the well. "Smart" wells with such equipment will be able to separate water out of the well stream so that it never goes to the surface. Instead a pump, controlled by a computer in the drill pipe, will inject the wastewater below the oil level.

Wading in Deeper

Perhaps the oil industry's last great frontier is in deep water, in fields that lie 1,000 meters or more below the surface of the sea. Petroleum at such depths used to be beyond reach, but no longer. Remotely controlled robot submarines can now install on the seafloor the complex equipment needed to guard against blowouts, to regulate the flow of oil at the prevailing high pressures and to prevent natural gas from freezing and plugging pipelines. Subsea complexes will link clusters of horizontal wells.

The collected oil will then be funneled both to tankers directly above and to existing platforms in shallower waters through long underwater pipelines. In just the next three years, such seafloor facilities are scheduled for construction in the Gulf of Mexico and off the shores of Norway, Brazil and West Africa.

More than deep water alone hinders the exploitation of offshore oil and gas fields. Large horizontal sheets of salt and basalt (an igneous rock) sometimes lie just underneath the seafloor in the deep waters of the continental margins. In conventional seismic surveys they scatter nearly all the sound energy so that oil fields below are hidden from view. But recently declassified U.S. Navy technology for measuring tiny variations in the force and direction of gravity, combined with ever expanding supercomputer capabilities, now allows geophysicists to see under these blankets of salt or basalt.

Extracting oil from beneath the deep ocean is still enormously expensive, but innovation and necessity have led to a new wave of exploration in that realm. Already the 10 largest oil companies working in deep water have discovered new fields that will add 5 percent to their combined oil reserves, an increase not yet reflected in global reserve estimates.

The technology for oil exploration and production will continue to march forward in the 21st century. Although it is unlikely that these techniques will entirely eliminate the impending shortfall in the supply of crude oil, they will buy critical time for making an orderly transition to a world fueled by other energy sources.

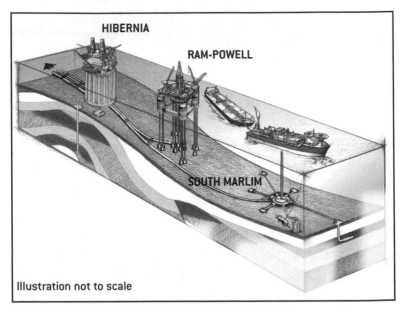

HIBERNIA

RAM-POWELL

SOUTH MARLIM

Illustration not to scale

Three new ways to tap oil fields that lie deep underwater have recently been developed. Hibernia (*left*), which began producing oil last November from a field in 80 meters of water off the coast of Newfoundland, Canada, took seven years and more than $4 trillion to construct. Its base, built from 450,000 tons of reinforced concrete is designed to withstand the impact of a million-ton iceberg. Hibernia is expected to recover 615 million barrels of oil over 18 years, using water and gas injection. Storage tanks will hold up to 1.3 million barrels of oil inside the base until it can be transferred to shuttle tankers. Most deepwater platforms send the oil back to shore through subsea pipelines. Ram-Powell platform (*center*), built by Shell Oil, Amoco and Exxon, began production in the Gulf of Mexico last September. The 46-story platform is anchored to 270-ton piles driven into the seafloor 980 meters below. Twelve tendons, each 71 centimeters in diameter, provide a strong enough mooring to withstand 22-meter waves and hurricane winds up to 225 kilometers per hour. The $1-billion facility can sink wells up to six kilometers into the seabed in order to tap the 125 million barrels of recoverable oil estimated to lie in the field. A 30-centimeter pipeline will transport the oil to platforms in shallower water 40 kilometers away. Ram-Powell is the third such tension leg platform completed by Shell in three years. Next year, Shell's plans call for an even larger platform, named Ursa, to start pumping 2.5 times as much oil as Ram-Powell from below 1,226 meters of water. Deepest oil well in active production (*below*) currently lies more than 1,709 meters beneath the waves of the South Atlantic Ocean, in the Marlim field off the coast of Campos, Brazil. The southern part of this field alone is thought to contain 10.6 billion barrels of oil. Such resources were out of reach until recently. Now remotely operated submarines are being used to construct production facilities on the sea bottom itself. The oil can then be piped to a shallower platform if one is nearby. Or, as in the case of the record-holding South Marlim 3B well, a ship can store the oil until shuttle tankers arrive. The challenge is to hold the ship steady above the well. Moorings can provide stability at depths up to about 1,500 meters. Beyond that limit, ships may have to use automatic thrusters linked to the Global Positioning System and beacons on the seafloor to actively maintain their position. These techniques may allow the industry to exploit oil fields under more than 3,000 meters of water in the near future.

The Author

Roger N. Anderson is director of petroleum technology research at the Energy Center of Columbia University. After growing up with a father in the oil industry, Anderson completed his Ph.D. in earth science at the Scripps Institution of Oceanography at the University of California, San Diego. He sits on the board of directors of Bell Geospace and 4-D Systems and spends his summers consulting for oil and service companies. Anderson has published more than 150 peer-reviewed scientific papers and holds seven U.S. patents.

Many people believe that natural gas is a promising potential solution to increasing problems with the oil supply. Because natural gas and fuels made from it are cleaner than other petroleum products, producing fewer pollutants upon combustion, they are also better for the environment than oil. The following article discusses how natural gas can be used as a starting material to make liquid fuel. This is a two-step process involving chemical reactions. The particular fuels produced depend on the temperature at which the second step takes place, but diesel is the most commonly produced fuel from this process.

In the first step of the process, methane, the primary component of natural gas, is converted into carbon monoxide and hydrogen, a mixture called syngas. This generally requires a lot of oxygen, which is expensive. Several groups of researchers are creating membranes to remove oxygen from air more easily and cheaply. A partnership between Air Products and several other companies and universities has developed commercial-size ceramic membranes for oxygen separation and is creating a prototype syngas *reactor with them. With new techniques like ceramic membranes, liquid fuels from natural gas will become cheaper and thus more competitive with other petroleum products. —LEH*

"Liquid Fuels from Natural Gas"
by Safaa A. Fouda
Scientific American, **March 1998**

Recently countless California motorists have begun contributing to a remarkable transition. Few of these drivers realize that they are doing something special when they tank up their diesel vehicles at the filling station. But, in fact, they are helping to wean America from crude oil by buying a fuel made in part from natural gas.

Diesel fuel produced in this unconventional way is on sale in California because the gas from which it is derived is largely free of sulfur, nitrogen and heavy metals—substances that leave the tailpipe as noxious pollutants. Blends of ordinary diesel fuel and diesel synthesized from natural gas (currently produced commercially by Shell in Indonesia) meet the toughest emissions standards imposed by the California Air Resources Board.

But natural gas is not only the cleanest of fossil fuels, it is also one of the most plentiful. Industry analysts estimate that the world holds enough readily recoverable natural gas to produce 500 billion barrels of synthetic crude—more than twice the amount of oil ever found in the U.S. Perhaps double that quantity of gas can be found in coal seams and in formations that release gas only slowly. Thus, liquid fuels derived from natural gas could keep overall production on the rise for about a decade after conventional supplies of crude oil begin to dwindle.

Although global stocks of natural gas are enormous, many of the deposits lie far from the people in need of energy. Yet sending gas over long distances often turns out to be prohibitively expensive. Natural gas costs four times as much as crude oil to transport through pipelines because it has a much lower energy density. The so-called stranded gas can be cooled and compressed into a liquid for shipping by tanker. Unfortunately, the conversion facilities required are large and complex, and because liquefied natural gas is hard to handle, the demand for it is rather limited.

But what if there were a cheap way to convert natural gas to a form that remains liquid at room temperature and pressure? Doing so would allow the energy to be piped to markets inexpensively. If the liquid happened to be a fuel that worked in existing vehicles, it could substitute for oil-based gasoline and diesel. And oil producers would stand to profit in many instances by selling liquid fuels or other valuable chemicals made using the gas coming from their wells.

Right now the gas released from oil wells in many parts of the world holds so little value that it is either burned on site or reinjected into the ground. In Alaska alone, oil companies pump about 200 million cubic meters (roughly seven billion cubic feet) of natural gas back into the ground daily—in large part to avoid burdening the atmosphere with additional carbon dioxide, a worrisome greenhouse gas.

But recent technical advances have prompted several oil companies to consider building plants to convert this natural gas into liquid form, which could then be delivered economically through the Alaska pipeline. On the Arabian Peninsula, the nation of Qatar is negotiating with three petrochemical companies to build gas conversion plants that would exploit a huge offshore field—a single reservoir that contains about a tenth of the world's proved gas reserves. And Norway's largest oil company, Statoil, is looking at building relatively small modules mounted on floating platforms to transform gas in remote North Sea fields into liquids. Although these efforts will use somewhat different technologies, they all must address the same fundamental problem in chemistry: making larger hydrocarbon molecules from smaller ones.

The Classic Formula

The main component of natural gas is methane, a simple molecule that has four hydrogen atoms neatly arrayed around one carbon atom. This symmetry makes methane particularly stable. Converting it to a liquid fuel requires first breaking its chemical bonds. High temperatures and pressures help to tear these bonds apart. So do cleverly designed catalysts, substances that can foster a chemical reaction without themselves being consumed.

The conventional "indirect" approach for converting natural gas to liquid form relies on brute force. First, the chemical bonds in methane are broken using steam, heat and a nickel-based catalyst to produce a mixture of carbon monoxide and hydrogen known as syngas (or, more formally, synthesis gas). This process is called steam re-forming.

The second step in the production of liquid fuels (or other valuable petrochemicals) from syngas uses a method invented in 1923 by Franz Fischer and Hans Tropsch. During World War II, Germany harnessed this technique to produce liquid fuels using syngas made from coal and atmospheric oxygen, thus establishing a reliable internal source for gasoline and diesel.

This Fischer-Tropsch technology has allowed Sasol in South Africa to produce liquid fuels commercially for decades using syngas derived from coal. The company uses the same basic technique today: syngas blown over a catalyst made of cobalt, nickel or iron transforms into various liquid hydrocarbons. Conveniently, the Fischer-Tropsch reaction gives off heat, and often this heat is used to drive the oxygen compressors needed to make syngas.

Just which liquids emerge from the reaction depends on temperature. For example, running a reaction vessel at 330 to 350 degrees Celsius (626 to 662 degrees Fahrenheit) will primarily produce gasoline and olefins (building blocks often used to make plastics). A cooler (180 to 250 degree C) operation will make predominantly diesel and waxes. In any case, a mixture results, so a third and final step is required to refine the products of the reaction into usable fuels.

Refining synthetic crudes derived from gas is in many respects easier than working with natural crude oil. Synthetic crude contains virtually no sulfur and has smaller amounts of cancer-causing compounds than are found in conventional oil. So the final products are premium-quality fuels that emit fewer harmful substances.

A Partial Solution

This brute-force method of converting gas to liquids is reliable, but it is expensive because it uses so much energy. Conventional steam re-forming compresses methane and water vapor to about 30 times normal atmospheric pressure and heats these reactants to about 900 degrees C. And one must add more heat still, to coax the energy-hungry reaction continuously along. This extra heat comes from injecting a small amount of oxygen into the mixture, which combusts some of the methane (and, as an added benefit, makes more syngas). Chemists call this latter maneuver partial oxidation.

In general, syngas is generated using various combinations of steam re-forming and partial oxidation. In most cases, the process requires large quantities of oxygen—and oxygen is costly. Existing methods of separating oxygen from air rely on refrigeration to cool and liquefy it, an energy-intensive and expensive manipulation. Hence, lowering the cost of oxygen is the key to making syngas cheaply.

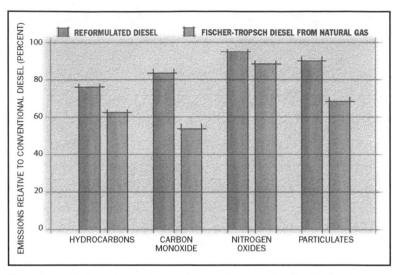

Harmful vehicle emissions were lowered somewhat in 1993, when U.S. regulations required that diesel fuel be reformulated to reduce pollution. Fuel derived from natural gas using Fischer-Tropsch synthesis creates even fewer emissions than reformulated diesel.

Fortunately, recent developments promise to revolutionize the way oxygen is produced over the next few years. One strategy is simply to work with air instead of pure oxygen. Syntroleum Corporation in Tulsa has developed a way to make liquid fuels using blown air and methane for the re-forming step, followed by Fischer-Tropsch synthesis. At sites where natural gas is sufficiently cheap (for example, places where it is now being flared), the process should prove profitable even at current crude oil prices. Together with Texaco and the English company Brown & Root, Syntroleum plans to build a commercial plant that will use this technique within two years.

Several other private companies, universities and government research laboratories are pursuing a wholly different approach to the oxygen problem: they are developing ceramic membranes through which only oxygen can pass. These membranes can then serve as filters to purify oxygen from air. Though still difficult and expensive to construct, laboratory versions work quite well. They should be commercially available within a decade.

Such materials could reduce the cost of making syngas by about 25 percent and lower the cost of producing liquid fuels by 15 percent. These savings would accrue because the production of syngas could be done at temperatures about 200 degrees lower than those currently used and because there would be no need to liquefy air. With cheap and plentiful oxygen, partial oxidation alone could supply syngas. This first step would then release energy rather than consume it.

My Canadian colleagues and I, along with researchers at the University of Florida, are now attempting to create a different kind of ceramic membrane that would offer yet another advantage. The membranes we are trying to develop would remove hydrogen from the gas mixture, driving the partial oxidation of methane forward and providing a stream of pure hydrogen that could be used later in refining the final products or as an energy source itself.

We also expect to see significant improvements soon in the catalysts used to make syngas. In particular, researchers at the University of Oxford are studying metal carbides, and my colleagues at the Canadian Center for Mineral and Energy Technology are investigating large pore zeolites. Both materials show great promise in reducing the soot generated during operation, a problem that not only plugs the reactor but also reduces the activity of the catalysts over time.

Cheaper than Oil?

Although the prospects for such brute-force methods of converting natural gas to liquid fuel improve every day, more ingenious techniques on the horizon would accomplish that transformation in a single step. This approach could potentially cut the cost of conversion in half, which would make liquid fuels produced from natural gas actually less expensive than similar products refined from crude oil.

Early efforts to achieve such "direct" conversion by using different catalysts and adding greater amounts of oxygen had produced mostly disappointment. The hydrocarbons that were formed proved more reactive than the methane supplied. In essence, they burned up faster than they were produced. Unless the product is somehow removed from the reaction zone, yields are too low to be practical.

Fortunately, researchers have recently found ways to circumvent this problem. The trick is to run the reaction at comparatively mild temperatures using exotic catalysts or to stabilize the product chemically— or to do both. For example, chemists at Pennsylvania State University have converted methane to methanol

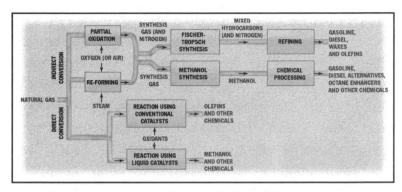

Chemical conversion of natural gas follows one of two general approaches. "Indirect" methods first form synthesis gas, or syngas (a mixture of carbon monoxide and hydrogen), by adding steam and oxygen or oxygen alone. (One company, Syntroleum, uses air instead, removing the unwanted nitrogen during later processing.) The second step synthesizes larger molecules from syngas, and the third step refines or chemically tailors the various products. "Direct" conversion of natural gas in one step requires an oxidant and may involve special liquid catalysts.

directly using a so-called homogeneous catalyst, a liquid that is thoroughly mixed with the reactants and held at temperatures lower than 100 degrees C. And Catalytica, a company in Mountain View, Calif., has achieved yields for direct conversion that are as high as 70 percent using a similar scheme. Its liquid catalyst creates a relatively stable chemical intermediate, methyl ester, that is protected from oxidation. The final product (a methanol derivative) is easily generated with one subsequent step.

Methanol (also known as wood alcohol) is valuable because it can be readily converted to gasoline or to an octane-boosting additive. And in the near future methanol (either used directly or transformed first into hydrogen gas) could also serve to power fuel-cell vehicles on a wide scale. Thus, methanol can be regarded as a convenient currency for storing and transporting energy.

Moreover, the reactions used to synthesize methanol can be readily adjusted to churn out diesel alternatives such as dimethyl ether, which produces far fewer troublesome pollutants when it burns. So far dimethyl ether, like propane, has found little use as a transportation fuel because it is a gas at room temperature and pressure. But recently Air Products, a supplier of industrial gases in Allentown, Pa., announced the production of a dimethyl ether derivative that is liquid at ambient conditions. So this substitute for conventional diesel fuel would reduce emissions without major changes to vehicles and fueling stations.

Now You're Cooking with Gas

Scientists and engineers are pursuing many other possible ways to improve the conversion of natural gas into liquids. For instance, process developers are constantly improving the vessels for the Fischer-Tropsch reaction to provide better control of heat and mixing.

The most ambitious efforts now under way attempt to mimic the chemical reactions used by specialized bacteria that consume methane in the presence of oxygen to produce methanol. Low temperature biological reactions of this kind are quite promising because they can produce specific chemicals using relatively little energy.

Whether or not this bold line of research ultimately succeeds, it is clear that even today natural gas can be converted into liquid fuels at prices that are only about 10 percent higher per barrel than crude oil. Modest improvements in technology, along with the improved economics that come from making specialty chemicals as well from gas, will broaden the exploitation of this abundant commodity in coming years. Such developments will also provide remarkably clean fuels—ones that can be easily blended with dirtier products refined from heavier crude oils to meet increasingly strict environmental standards. So the benefits to society will surely multiply as people come to realize that natural gas can do much more than just run the kitchen stove.

The Author

Safaa A. Fouda received a doctorate in chemical engineering from the University of Waterloo in 1976. Since 1981 she has worked at the CANMET Energy Technology Center, a Canadian government laboratory in Nepean, Ontario. There she manages a group of researchers studying natural gas conversion, emissions control, waste oil recycling and liquid fuels from renewable sources. Recently she headed an international industrial consortium intent on developing better methods to convert natural gas to liquid fuels.

2. Nuclear Energy

Currently, thirty-one countries together contain 441 operational nuclear reactors that produce 363 billion watts of power annually, according to the World Nuclear Association. Most of these are referred to as Generation II nuclear power plants: they are the large central station nuclear plants that evolved from experimental prototypes (Generation I). These Generation II plants were built in the 1970s and 1980s during a flurry of construction, after which building of new plants slowed, especially following the serious reactor accidents at Three Mile Island in Middletown, Pennsylvania, in 1979 and Chernobyl in the Ukraine in 1986.

Recently, construction of nuclear power plants began to increase again, particularly in Asia. Many Generation II plants are due to be decommissioned in the next decade, which will increase construction even more. Some Generation III plants—advanced reactors with more safety features—have been built, and more are on the way. But researchers are already looking ahead to Generation IV systems, as discussed in the following article. Expected to be operational by 2030, these plants will incorporate new technologies that are intended to be safer and more efficient.

One advantage of some of these new technologies is that they will operate at such high temperatures that they will produce hydrogen from water. Currently, 97 percent of hydrogen is produced from fossil fuels. Production of hydrogen as a by-product of nuclear power could make fuel-cell cars an even greener technology. —LEH

"Next-Generation Nuclear Power"
by James A. Lake, Ralph G. Bennett, and John F. Kotek
Scientific American, January 2002

Rising electricity prices and last summer's rolling blackouts in California have focused fresh attention on nuclear power's key role in keeping America's lights on. Today 103 nuclear plants crank out a fifth of the nation's total electrical output. And despite residual public misgivings over Three Mile Island and Chernobyl, the industry has learned its lessons and established a solid safety record during the past decade. Meanwhile the efficiency and reliability of nuclear plants have climbed to record levels. Now with the ongoing debate about reducing greenhouse gases to avoid the potential onset of global warming, more people are recognizing that nuclear reactors produce electricity without discharging into the air carbon dioxide or pollutants such as nitrogen oxides and smog-causing sulfur compounds. The world demand for energy is projected to rise by about 50 percent by 2030 and to nearly double by 2050. Clearly, the time seems right to reconsider the future of nuclear power.

No new nuclear plant has been ordered in the U.S. since 1978, nor has a plant been finished since 1995. Resumption of large-scale nuclear plant construction requires that challenging questions be addressed regarding the achievement of economic viability, improved operating safety, efficient waste management and resource utilization, as well as weapons nonproliferation, all of which are influenced by the design of the nuclear reactor system that is chosen.

Designers of new nuclear systems are adopting novel approaches in the attempt to attain success. First, they are embracing a system-wide view of the nuclear fuel cycle that encompasses all steps from the mining of ore through the management of wastes and the development of the infrastructure to support these steps. Second, they are evaluating systems in terms of their sustainability— meeting present needs without jeopardizing the ability of future generations to prosper. It is a strategy that helps to illuminate the relation between energy supplies and the needs of the environment and society. This emphasis on sustainability can lead to the development of nuclear energy-derived products besides electrical power, such as hydrogen fuel for transportation. It also promotes the exploration of alternative reactor designs and nuclear fuel-recycling processes that could yield significant reductions in waste while recovering more of the energy contained in uranium.

We believe that wide-scale deployment of nuclear power technology offers substantial advantages over other energy sources yet faces significant challenges regarding the best way to make it fit into the future.

Future Nuclear Systems

In response to the difficulties in achieving sustainability, a sufficiently high degree of safety and a competitive economic basis for nuclear power, the U.S. Department of Energy initiated the Generation IV program in 1999. Generation IV refers to the broad division of nuclear designs into four categories: early prototype reactors (Generation I), the large central station nuclear power plants of today (Generation II), the advanced lightwater reactors and other systems with inherent safety features that have been designed in recent years (Generation III), and the next-generation systems to be designed and built two decades from now (Generation IV). By 2000 international interest in the Generation IV project had resulted in a nine-country coalition that includes Argentina, Brazil, Canada, France, Japan, South Africa, South Korea, the U.K. and the U.S. Participating states are mapping out and collaborating on the research and development of future nuclear energy systems.

Although the Generation IV program is exploring a wide variety of new systems, a few examples serve to illustrate the broad approaches reactor designers

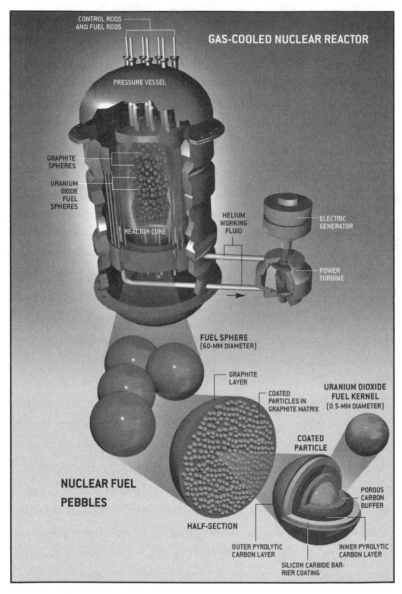

GAS-COOLED NUCLEAR REACTOR

CONTROL RODS AND FUEL RODS

PRESSURE VESSEL

GRAPHITE SPHERES

URANIUM OXIDE FUEL SPHERES

REACTOR CORE

HELIUM WORKING FLUID

ELECTRIC GENERATOR

POWER TURBINE

FUEL SPHERE (60-MM DIAMETER)

GRAPHITE LAYER

COATED PARTICLES IN GRAPHITE MATRIX

URANIUM DIOXIDE FUEL KERNEL (0.5-MM DIAMETER)

COATED PARTICLE

NUCLEAR FUEL PEBBLES

POROUS CARBON BUFFER

HALF-SECTION

OUTER PYROLYTIC CARBON LAYER

INNER PYROLYTIC CARBON LAYER

SILICON CARBIDE BARRIER COATING

Nuclear fuel pebbles: Round fuel elements, which permit continuous refueling during operation, cannot melt and degrade only slowly, providing a substantial safety margin.

Gas-cooled nuclear reactor: Core of a pebble-bed nuclear reactor (*shown in concept form, facing page*) contains hundreds of thousands of pebbles—spherical uranium oxide fuel and graphite elements. This innovative design offers significantly higher thermal efficiencies than current light-water reactors do.

are developing to meet their objectives. These next-generation systems are based on three general classes of reactors: gas-cooled, water-cooled and fast-spectrum.

Gas-Cooled Reactors

Nuclear reactors using gas (usually helium or carbon dioxide) as a core coolant have been built and operated successfully but have achieved only limited use to date. An especially exciting prospect known as the pebble-bed modular reactor possesses many design features that go a good way toward meeting Generation IV goals. This gas-cooled system is being pursued by engineering teams in China, South Africa and the U.S. South Africa plans to build a full-size prototype and begin operation in 2006.

The pebble-bed reactor design is based on a fundamental fuel element, called a pebble, that is a billiard-ball-size graphite sphere containing about 15,000 uranium oxide particles with the diameter of poppy seeds [*see illustration at left*]. The evenly dispersed particles each have several high-density coatings on them. One of the layers, composed of tough silicon carbide ceramic, serves as a pressure vessel to retain the products of nuclear fission during reactor operation or accidental temperature excursions. About 330,000 of these spherical fuel pebbles are placed into a metal vessel surrounded by a shield of graphite blocks. In addition, as many as 100,000 unfueled graphite pebbles are loaded into the core to shape its power and temperature distribution by spacing out the hot fuel pebbles.

Heat-resistant refractory materials are used throughout the core to allow the pebble-bed system to operate much hotter than the 300 degree Celsius temperatures typically produced in today's light-water-cooled (Generation II) designs. The helium working fluid, exiting the core at 900 degrees C, is fed directly into a gas turbine/generator system that generates electricity at a comparatively high 40 percent thermal efficiency level, one quarter better than current light-water reactors.

The comparatively small size and the general simplicity of pebble-bed reactor designs add to their economic feasibility. Each power module, producing 120 megawatts of electrical output, can be deployed in a unit one tenth the size of today's central station plants, which permits the development of more flexible, modest-scale projects that may offer more favorable economic results. For example, modular systems can be manufactured in the factory and then shipped to the construction site.

The pebble-bed system's relative simplicity compared with current designs is dramatic: these units have only about two dozen major plant subsystems, compared with about 200 in light-water reactors. Significantly, the operation of these plants can be extended into a temperature range that makes possible the low emissions production of hydrogen from water or other feedstocks for use in fuel cells and clean-burning transportation engines, technologies on which a sustainable hydrogen-based energy economy could be based.

These next-generation reactors incorporate several important safety features as well. Being a noble gas, the helium coolant will not react with other materials, even at high temperatures. Further, because the fuel elements and reactor core are made of refractory materials, they cannot melt and will degrade only at the extremely high temperatures encountered in accidents (more than 1,600 degrees C), a characteristic that affords a considerable margin of operating safety.

Yet other safety benefits accrue from the continuous, on-line fashion in which the core is refueled: during operation, one pebble is removed from the bottom of the core about once a minute as a replacement is placed on top. In this way, all the pebbles gradually move down through the core like gumballs in a dispensing machine, taking about six months to do so. This feature means that the system contains the optimum amount of fuel for operation, with little extra fissile reactivity. It eliminates an entire class of excess-reactivity accidents that can occur in current water-cooled reactors. Also, the steady movement of pebbles through regions of high and low power production means that each experiences less extreme operating conditions on average than do fixed fuel configurations, again adding to the unit's safety margin. After use, the spent pebbles must be placed in long-term storage repositories, the same way that used-up fuel rods are handled today.

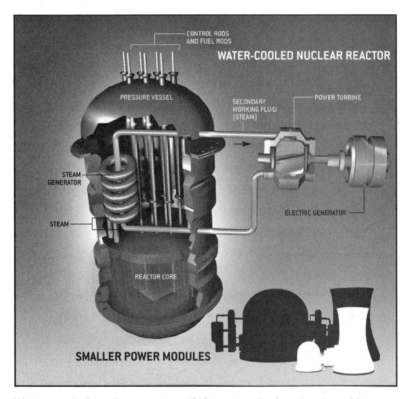

Water-cooled nuclear reactor: IRIS reactor design developed by Westinghouse Electric (*depicted in conceptual form*) is novel in that both the steam generator (heat exchanger) and the control rod actuator drives are enclosed within the thick steel pressure vessel.

Small power modules: Containment buildings for the compact IRIS reactor can be reduced in size. The reactor's lower power output, ranging from 100 to 350 megawatts, can make these units more economical as well.

Water-Cooled Reactors

Even standard water-cooled nuclear reactor technology has a new look for the future. Aiming to overcome the possibility of accidents resulting from loss of coolant (which occurred at Three Mile Island) and to simplify the overall plant, a novel class of Generation IV systems has arisen in which all the primary components are contained in a single vessel. An American design in this class is the international reactor innovative and secure (IRIS) concept developed by Westinghouse Electric.

Housing the entire coolant system inside a damage-resistant pressure vessel means that the primary system cannot suffer a major loss of coolant even if one of its large pipes breaks. Because the pressure vessel will not allow fluids to escape, any resulting accident is limited to a much more moderate drop in pressure than could occur in previous designs.

To accomplish this compact configuration, several important simplifications are incorporated in these reactors. The subsystems within the vessel are stacked to enable passive heat transfer by natural circulation during accidents. In addition, the control rod drives are located in the vessel, eliminating the chance that they could be ejected from the core. These units can also be built as small power modules, thereby allowing more flexible and lower-cost deployment.

Designers of these reactors are also exploring the potential of operating plants at high temperature and pressure (more than 374 degrees C and 221 atmospheres), a condition known as the critical point of water, at which the distinction between liquid and vapor blurs. Beyond its critical point, water behaves as a continuous fluid with exceptional specific heat (thermal storage capacity) and superior heat transfer (thermal conductance) properties. It also does not boil as it heats up or flash to steam if it undergoes rapid depressurization. The primary advantage to operating above the critical point is that the system's thermal efficiency can reach as high as 45 percent and approach the elevated temperature regime at which hydrogen fuel production can become viable.

Although reactors based on supercritical water appear very similar to standard Generation II designs at first glance, the differences are many. For instance, the cores of the former are considerably smaller, which helps to economize on the pressure vessel and the surrounding plant. Next, the associated steam-cycle equipment is substantially simplified because it operates with a single-phase working fluid. In addition, the smaller core and the low coolant density reduce the volume of water that must be held within the containment vessel in the event of an accident. Because the low-density coolant does not moderate the energy of the neutrons, fast-spectrum reactor designs, with their associated sustainability benefits, can be contemplated. The chief downside to supercritical water systems is that the coolant becomes increasingly corrosive. This means that new materials and methods to control corrosion and erosion must be developed. Supercritical water reactor research is ongoing in Canada, France, Japan, South Korea and the U.S.

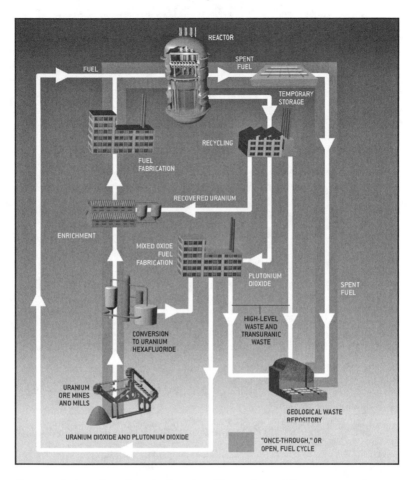

Open and closed nuclear fuel cycles: "Once-through," or open, nuclear fuel cycle (*shown in gray*) takes uranium ore, processes it into fissile fuel, burns it a single time in a reactor and then disposes of it in a geological repository. This approach, which is employed in the U.S., uses only 1 percent of the uranium's energy content. In a closed cycle (*shown in white*), the spent fuel is processed to reclaim its uranium and plutonium fuel content for reuse. This recycling method is used today in France, Japan and the U.K. Future closed cycles based on fast-spectrum reactors could reclaim other actinides that are currently treated as waste.

Fast-Spectrum Reactors

A design approach for the longer term is the fast-spectrum (or high-energy neutron) reactor, another type of Generation IV system. An example of this class of reactor is being pursued by design teams in France, Japan, Russia, South Korea and elsewhere. The American fast-reactor development program was canceled in 1995, but U.S. interest might be revived under the Generation IV initiative.

Most nuclear reactors employ a thermal, or relatively low energy, neutronemissions spectrum. In a thermal reactor the fast (high-energy) neutrons generated in the fission reaction are slowed down to "thermal" energy levels as they collide with the hydrogen in water or other light nuclides. Although these reactors are economical for generating electricity, they are not very effective in producing nuclear fuel (in breeder reactors) or recycling it.

Most fast-spectrum reactors built to date have used liquid sodium as the coolant. Future versions of this reactor class may utilize sodium, lead, a lead-bismuth alloy or inert gases such as helium or carbon dioxide. The higher-energy neutrons in a fast reactor can be used to make new fuel or to destroy long-lived wastes from thermal reactors and plutonium from dismantled weapons. By recycling the fuel from fast reactors, they can deliver much more energy from uranium while reducing the amount of waste that must be disposed of for the long term. These breeder-reactor designs are one

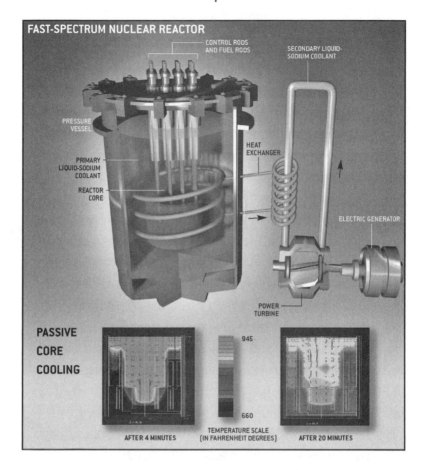

FAST-SPECTRUM NUCLEAR REACTOR

CONTROL RODS AND FUEL RODS

SECONDARY LIQUID-SODIUM COOLANT

PRESSURE VESSEL

HEAT EXCHANGER

PRIMARY LIQUID-SODIUM COOLANT

REACTOR CORE

ELECTRIC GENERATOR

POWER TURBINE

PASSIVE CORE COOLING

945

660

AFTER 4 MINUTES

TEMPERATURE SCALE (IN FAHRENHEIT DEGREES)

AFTER 20 MINUTES

Fast-spectrum nuclear reaction: Cores of fast-spectrum nuclear reactors such as General Electric's Super PRISM design (*shown in conceptual form*), which produce fast (high-energy) neutrons, are often cooled with molten metals. In breeder-reactor configurations, these high-energy neutrons are used to create nuclear fuel.

Passive core cooling: Temperature distributions show how the high heat-transfer properties of liquid-metal coolants can lower the reactor core temperature passively following the accidental loss of the external heat sink.

of the keys to increasing the sustainability of future nuclear energy systems, especially if the use of nuclear energy is to grow significantly.

Beyond supporting the use of a fastneutron spectrum, metal coolants have several attractive qualities. First, they possess exceptional heat-transfer properties, which allows metal-cooled reactors to withstand accidents like the ones that happened at Three Mile Island and Chernobyl. Second, some (but not all) liquid metals are considerably less corrosive to components than water is, thereby extending the operating life of reactor vessels and other critical subsystems. Third, these high-temperature systems can operate near atmospheric pressure, greatly simplifying system design and reducing potential industrial hazards in the plant.

More than a dozen sodium-cooled reactors have been operated around the world. This experience has called attention to two principal difficulties that must be overcome. Sodium reacts with water to generate high heat, a possible accident source. This characteristic has led sodium-cooled reactor designers to include a secondary sodium system to isolate the primary coolant in the reactor core from the water in the electricity-producing steam system. Some new designs focus on novel heat-exchanger technologies that guard against leaks.

The second challenge concerns economics. Because sodium-cooled reactors require two heat-transfer steps between the core and the turbine, capital costs are increased and thermal efficiencies are lower than those of the most advanced gas- and water-cooled concepts (about 38 percent in an advanced sodium-cooled reactor compared with 45 percent in a supercritical water reactor). Moreover, liquid metals are opaque, making inspection and maintenance of components more difficult.

Next-generation fast-spectrum reactor designs attempt to capitalize on the advantages of earlier configurations while addressing their shortcomings. The technology has advanced to the point at which it is possible to envision fast-spectrum reactors that engineers believe will pose little chance of a meltdown. Further, nonreactive coolants such as inert gases, lead or lead-bismuth alloys may eliminate the need for a secondary coolant system and improve the approach's economic viability.

Nuclear energy has arrived at a crucial stage in its development. The economic success of the current generation of plants in the U.S. has been based on improved management techniques and careful practices, leading to growing interest in the purchase of new plants. Novel reactor designs can dramatically improve the safety, sustainability and economics of nuclear energy systems in the long term, opening the way to their widespread deployment.

The Authors

James A. Lake, Ralph G. Bennett, and John F. Kotek play leading roles in the U.S. nuclear energy program. Lake is associate laboratory director for nuclear and energy systems

*at the U.S. Department of Energy's Idaho National
Engineering and Environmental Laboratory (INEEL),
where he heads up research and development programs on
nuclear energy and safety as well as renewable and fossil
energy. In 2001 he served as president of the American
Nuclear Society. Bennett is director of nuclear energy at
INEEL and a member of the team that leads the DOE's
Generation IV effort. Kotek is manager of the special
projects section at Argonne National Laboratory–West in
Idaho and a member of the team that directs the DOE's
Generation IV effort. Before joining Argonne in 1999, he
was associate director for technology in the DOE's Office
of Nuclear Energy, Science and Technology.*

*Most atoms are stable, but some are radioactive,
meaning that they emit radiation that can cause
illness and even death in living organisms. The
half-life of a radioactive element is the time that
it takes for one half of a sample to decay and
become nonradioactive. In a nuclear reactor,
chain reactions between radioactive elements
produce energy. Radioactive materials remain as
a by-product of the reaction, however; some of
these have half-lives of hundreds, thousands, or
even millions of years.*

*Many people argue that nuclear power is a
cleaner alternative to fossil fuels. Nuclear plants*

*produce energy without emitting air pollutants
or greenhouse gases, and they do not use non-
renewable petroleum products. The amount of
nuclear waste produced, on the other hand, is
problematic. According to the World Nuclear
Association, nuclear plants produce about
12,000 metric tons of high-level waste—the
most radioactive type—each year.*

*The following article provides an introduction
to the various strategies used to dispose of
nuclear waste. All have their pros and cons, and
none has seen full implementation. —LEH*

"Disposing of Nuclear Waste"
by the Editors
Scientific American, September 1995

At 3:49 PM on December 2, 1942, in a converted
squash court under the football stands at the University
of Chicago, a physicist slid back some control rods in
the first nuclear reactor and ushered in a new age.
Four and a half minutes later the world had its first
nuclear waste. Since then, mountains of high-level
waste have joined that original molehill in Chicago:
according to the International Atomic Energy Agency,
about 10,000 cubic meters of high-level waste accumulate each year.

This massive amount of radioactive material has
no permanent home. Not a single country has managed
to implement a long-term plan for storing it; each

relies on interim measures. In the U.S., for example, used fuel rods are generally kept in pools near a reactor until they are cool enough for dry storage in steel casks elsewhere at the site.

However large or small a role nuclear fission plays in meeting future energy needs, safely disposing of intensely radioactive by-products will remain a top priority, if only because so much material already exists. Among the most promising technologies are:

- **Permanent subterranean storage.** All the countries that have significant amounts of high-level waste are currently hoping to store it deep underground in geologically stable areas. In the U.S. version of this plan, spent fuel rods are to be sealed in steel canisters and allowed to cool aboveground for several years. If the high-level waste is in liquid form, it will be dried and "vitrified," or enclosed in glass logs, before being put in canisters. These containers will later be placed in canisters up to 5.6 meters long, which will, in turn, be inserted in holes drilled in the rock floor of warehouselike caverns, hundreds of meters below the surface. The holes will be covered with plugs designed to shield the room above from radiation. Federal officials had hoped to build the U.S. repository beneath Yucca Mountain in Nevada, but local authorities are fighting to keep the repository out of the state.

- **Entombment under the seabed.** Pointed canisters containing the waste could be dropped from ships to the floor far below, where they would penetrate and embed themselves tens of meters down. The advantages stem from the ability to use seafloor sites that are stable and remote from the continents. In a variation on this idea, the canisters could be dropped into deep ocean trenches, where they would be pulled into the earth's mantle by the geologic process of subduction.

  Considered the most scientifically sound proposals by some experts, these—as with all schemes involving the oceans—have been rejected by science policymakers because of concerns over the potentially adverse public reaction and the plan's possible violation of international treaties barring the disposal of radioactive waste at sea.

- **Nuclear transmutation.** The troublesome components of high-level waste are a relatively small number of materials that are radioactive for tens of thousands of years. If appropriately bombarded with neutrons, however, the materials can be transmuted into others that are radioactive for only hundreds or possibly just tens of years. A repository would still be needed. It could hold much more material, though, because the amount of heat emitted by the waste would be significantly reduced.

A very small scale form of transmutation has been carried out for decades in specially designed experimental reactors. Lately scientists at Los Alamos National Laboratory have proposed using a high-energy accelerator to make the process more rapid and efficient. According to Wendell D. Weart, a senior scientist at Sandia National Laboratories, the main challenge would be concentrating the nuclear materials. "No one's ever tried to do it with the degree of separation that this would require," he says. "It's not easy to work with intensely radioactive systems, with this degree of chemical separation. It would be a neat trick."

Some less highly regarded proposals are:

- **Shooting nuclear waste into space or the sun.** This idea has been rejected because of its enormous expense, as well as the possibility that a loaded rocket could blow up before leaving the earth's atmosphere.

- **Storage under a polar icecap.** High-level waste generates enough heat to melt not only ice but possibly even rock. Perhaps because of that characteristic, the idea has not won many converts.

- **Dissolving waste in the world's oceans.**
  Uniformly spread over much of the surface of
  the earth, the radioactivity would be small
  compared with the background level, proponents
  have insisted.

---

*As discussed in the previous article, radioactive
waste is a growing problem for the United States
and other nations. As of 2003, the United States
had accumulated about 49,000 metric tons of
spent nuclear fuel and 22,000 canisters of
defense-related nuclear waste, according to the
Department of Energy.*

*The United States intends to build a huge
subterranean repository for nuclear waste at
Yucca Mountain in Nevada. The facility has
encountered many roadblocks, however, including
fierce local and statewide opposition, and its
opening has been delayed. Regardless, Yucca
Mountain's intended capacity is only 70,000
metric tons of radioactive waste, whereas the
Department of Energy predicts that the United
States will have amassed 135,000 metric tons
of waste by 2035. Clearly, other repositories or
disposal methods will be needed to augment the
Yucca repository—if, indeed, it ever opens.*

No one wants a radioactive waste facility in his or her backyard, which is one of the hurdles Yucca Mountain has had to face. That is why burying radioactive waste under the seabed, far from human habitation, is an attractive option to many. The following article discusses how this could feasibly be carried out. With most of the government and industry's focus being placed on land-based disposal, however, research on subseabed disposal has been scarce in recent years. —LEH

"Burial of Radioactive Waste Under the Seabed"
by Charles D. Hollister and Steven Nadis
Scientific American, January 1998

On the floor of the deep oceans, poised in the middle of the larger tectonic plates, lie vast mud flats that might appear, at first glance, to constitute some of the least valuable real estate on the planet. The rocky crust underlying these "abyssal plains" is blanketed by a sedimentary layer, hundreds of meters thick, composed of clays that resemble dark chocolate and have the consistency of peanut butter. Bereft of plant life and sparsely populated with fauna, these regions are relatively unproductive from a biological standpoint and largely devoid of mineral wealth.

Yet they may prove to be of tremendous worth, offering a solution to two problems that have bedeviled humankind since the dawn of the nuclear age: these neglected sub-oceanic formations might provide a permanent resting place for high-level radioactive wastes and a burial ground for the radioactive materials removed from nuclear bombs. Although the disposal of radioactive wastes and the sequestering of material from nuclear weapons pose different challenges and exigencies, the two tasks could have a common solution: burial below the seabed.

High-level radioactive wastes—in the form of spent fuel rods packed into pools at commercial nuclear power plants or as toxic slurries housed in tanks and drums at various facilities built for the production of nuclear weapons—have been accumulating for more than half a century, with no permanent disposal method yet demonstrated. For instance, in the U.S. there are now more than 30,000 metric tons of spent fuel stored at nuclear power plants, and the amount grows by about 2,000 metric tons a year. With the nuclear waste repository under development at Yucca Mountain, Nev., now mired in controversy and not expected to open before 2015 at the earliest [see "Can Nuclear Waste Be Stored Safely at Yucca Mountain?" by Chris G. Whipple; SCIENTIFIC AMERICAN, June 1996], pressure is mounting to put this material somewhere.

The disposition of excess plutonium and uranium taken from decommissioned nuclear weapons is an even more pressing issue, given the crisis that might ensue if such material were to fall into the wrong hands. The U.S. and Russia have each accumulated more than

100 metric tons of weapons-grade plutonium, and each country should have at least 50 metric tons of excess plutonium, plus hundreds of tons of highly enriched uranium, left over from dismantled nuclear weapons. Preventing terrorists or "rogue states" from acquiring this material is, obviously, a grave concern, given that a metric ton of plutonium could be used to make hundreds of warheads, the precise number depending on the size of the bomb and the ingenuity of the designer.

The Clinton administration has endorsed two separate methods for ridding the nation of this dangerous legacy. Both entail significant technical, economic and political uncertainties. One scheme calls for the surplus weapons plutonium to be mixed with radioactive wastes and molded into a special type of glass (a process called vitrification) or, perhaps, ceramic for subsequent burial at a site yet to be chosen. The glass or ceramic would immobilize the radioactive atoms (to prevent them from seeping into the surrounding environment) and would make deliberate extraction of the plutonium difficult. But the matrix material does not shield against the radiation, so vitrified wastes would still remain quite hazardous before disposal. Moving ahead with vitrification in the U.S. has required construction of a new processing plant, situated near Aiken, S.C. Assuming this facility performs at its intended capacity, each day it will produce just one modest cylinder of glass containing about 20 or so kilograms of plutonium. The projected cost is $1.4 million for each of these glassy logs. And after that considerable expense and effort, someone still has to dispose of the highly radioactive products of this elaborate factory.

The second option would be to combine the recovered plutonium with uranium oxide to create a "mixed oxide" fuel for commercial reactors—although most nuclear power plants in the U.S. would require substantial modification before they could run on such a blend. This alternative measure of consuming mixed-oxide fuels at commercial power plants is technically feasible but nonetheless controversial. Such activities would blur the traditional separation between military and civilian nuclear programs and demand heightened security, particularly at

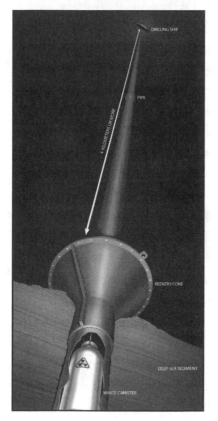

Steel pipe, lowered from a ship on the surface, would be used to drill holes in the deep-sea muds and, later, convey nuclear waste containers for permanent burial-according to the plan envisioned. Mud pumped into the borehole would then seal the nuclear refuse within the clay-rich undersea formation, effectively isolating the radioactive materials.

mixed-oxide fabrication plants (of which none currently exist in the U.S.), where material suitable for building a nuclear bomb might be stolen. And in the end, mixed-oxide reactors would produce other types of radioactive waste. Hence, neither of the schemes planned for disposing of material from nuclear weapons is entirely satisfactory.

Pressing Problems

For the past 15 years, the operators of nuclear power plants in the U.S. have been paying the Department of Energy in advance for the eventual storage or disposal of their wastes. Even though there is no place yet available to put this radioactive refuse, the courts have ordered the DOE to meet its contractual obligations and begin accepting expended fuel rods from nuclear utilities this year. It is not at all clear what the DOE will do with these materials. One plan supported by the U.S. Senate is to build a temporary storage facility in Nevada near the Yucca Mountain site, but President Bill Clinton opposes this stopgap measure. In any event, the mounting pressure to take some action increases the likelihood of hasty, ill-considered judgments. The best course, in our opinion, would be to do nothing too drastic for now; immediate action should be limited to putting the spent fuel currently residing in cooling ponds into dry storage as needed and trying to stabilize the leaks in high-level-waste containers at weapons sites, while scientists and engineers thoroughly investigate all reasonable means for permanent disposal.

Although some ambitious thinkers have suggested that nuclear waste might one day be launched into space and from there cast into the sun, most people who have studied the problem agree that safety and economy demand that the waste be put permanently underground. Curiously, the search for a suitable nuclear graveyard has been confined almost exclusively to sites on the continents, despite the fact that geologic formations below the world's oceans, which cover some 70 percent of the planet's surface, may offer even greater potential. The disposal of nuclear weapons and wastes below the seabed should not be confused with disposal in the deep-ocean trenches formed at the juncture of two tectonic plates—a risky proposition that would involve depositing waste canisters into some of the most geologically unpredictable places on the earth, with great uncertainty as to where the material would finally reside.

Subseabed disposal, in contrast, would utilize some of the world's most stable and predictable terrain, with radioactive waste or nuclear materials from warheads "surgically" implanted in the middle of oceanic tectonic plates. Selecting sites for disposal that are far from plate boundaries would minimize chances of disruption by volcanoes, earthquakes, crustal shifts and other seismic activity. Many studies by marine scientists have identified broad zones in the Atlantic and Pacific that have remained geologically inert for tens of millions of years. What is more, the clay-rich muds that would entomb the radioactive materials have intrinsically favorable characteristics: low permeability to water, a high adsorption capacity for these dangerous elements and a natural plasticity that enables the ooze to seal up any cracks or rifts that might develop around a waste container. So the exact form of the wastes (for example, whether they are vitrified or not) does not affect the feasibility of this approach. No geologic formations on land are known to offer all these favorable properties.

It is also important to note that disposal would not be in the oceans, per se, but rather in the sediments below. Placing nuclear waste canisters hundreds of meters underneath the floor of the deep ocean (which is, itself, some five or so kilometers below the sea surface) could be accomplished using standard deep-sea drilling techniques. The next step—backfilling to seal and pack the boreholes—is also a routine practice. This technology has proved itself through decades of use by the petroleum industry to probe the continental shelves and, more recently, by members of the Ocean Drilling Program, an international consortium of scientific researchers, to explore deeper locales.

We envision a specialized team of drillers creating boreholes in the abyssal muds and clays at carefully selected locations. These cylindrical shafts, some tens to hundreds of meters deep, would probably be spaced several hundred meters apart to allow for easy maneuvering. Individual canisters, housing plutonium or other radioactive wastes, would then be lowered by cable into the holes. The canisters would be stacked vertically but separated by 20 or more meters of mud, which could be pumped into the hole after each canister was emplaced.

As is the case for disposal within Yucca Mountain, the waste canisters themselves would last a few thousand years at most. Under the seabed, however, the muddy clays, which cling tenaciously to plutonium and many other radioactive elements, would prevent these substances from seeping into the waters above. Experiments conducted as part of an international research program concluded that plutonium (and other transuranic elements) buried in the clays would not migrate more than a few meters from a breached canister after even 100,000 years. The rates of migration for uranium and some other radioactive waste elements need yet to be properly determined. Still, their burial several tens to 100 meters or more into the sediments would most likely buy enough time for the radioactivity of all the waste either to decay or to dissipate to levels below those found naturally in seawater.

The Seabed Working Group, as the now defunct research program was called, consisted of 200 investigators from 10 countries. Led by the U.S. and sponsored by the Nuclear Energy Agency of the Organization for Economic Cooperation and Development, the project ran from 1976 to 1986 at a total cost of about $120 million. This program was an outgrowth of a smaller effort at Sandia National Laboratories that was initiated in response to a suggestion by one of the authors (Hollister), who conceived of the idea of subseabed disposal in 1973.

As part of the international program, scientists extracted core samples of the seabed and made preliminary environmental observations at about half a dozen sites in the northern Atlantic and Pacific oceans. The collected sediments showed an uninterrupted history of geologic tranquillity over the past 50 to 100 million years. And there is no reason to believe that these particular sites are extraordinary. On the contrary, thousands of cores from other midplate locations since examined as part of the Ocean Drilling Program indicate that the sediments that were studied originally are typical of the abyssal clays that cover nearly 20 percent of the earth. So one thing is clear: although other factors may militate against subseabed disposal, it will not be constrained by a lack of space.

Reviving an Old Idea

The Seabed Working Group concluded that although a substantial body of information supports the technical feasibility of the concept, further research "should be conducted before any attempt is made to use seabed disposal for high-level waste and spent fuel." Unfortunately, the additional investigations were never carried out because the U.S.—the principal financial backer of this research—cut off all funding in 1986 so that the nation could concentrate its efforts on land-based disposal. A year later the federal government elected to focus exclusively on developing a repository at Yucca Mountain—a shortsighted decision, especially in view of current doubts as to whether the facility will ever open.

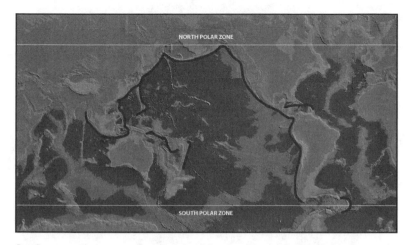

Seafloor provinces are not all suited for the disposal of nuclear wastes. In searching for candidate areas, scientists would probably eliminate places where the ocean floor is shallower than about four kilometers (*light gray water*), because these areas coincide with plate-tectonic spreading centers and are often blanketed by inappropriate types of sediments. They would also rule out other regions of tectonic activity, such as plate collision (*dark gray coastal lines*) or vulcanism. Polar zones (latitudes higher than 60 degrees) would be discounted because marine sediments there commonly contain coarse rock fragments carried in by icebergs. Even after these and other broad areas (such as around continental rises, where the sediments are thick enough to house valuable quantities of oil or gas) are exempted, vast stretches of seafloor still offer ample possibilities for burying nuclear wastes (*dark gray water*).

And even if the Yucca Mountain repository does become operational, it will not be able to handle all the high-level wastes from military and commercial sources that will have accumulated by the time of its inauguration, let alone the 2,000 or more tons of waste each year the nuclear industry will continue to churn out.

At some point, policymakers are going to have to face this reality and start exploring alternative sites and approaches. This view was precisely the conclusion expressed in a 1990 report from the National Academy of Sciences, which said that alternatives to mined geologic repositories, including subseabed disposal, should be pursued—a recommendation that remains absolutely valid today.

Fortunately, most of the experiments needed to assess more fully both the reliability and safety of subseabed disposal have been designed, and in many cases prototype equipment has already been built. One important experiment that remains to be done would be to test whether plutonium and other radioactive elements move through ocean-floor clays at the same rates measured in the laboratory. And more work is required to learn how the heat given off by fuel rods (caused by the rapid decay of various products of nuclear fission) would affect surrounding clays.

Research is also needed to determine the potential for disturbing the ecology of the ocean floor and the waters above. At present, the evidence suggests that mobile, multi-cellular life-forms inhabit only the top meter or so of the abyssal clays. Below a meter, there appear to be no organisms capable of transporting radioactive substances upward to the seafloor. Still, scientists would want to know exactly what the consequences would be if radioactive substances diffused to the seafloor on their own. Researchers would want to ascertain, for instance, exactly how quickly relatively soluble carriers of radioactivity (such as certain forms of cesium and technetium) would be diluted to background levels. And they would want to be able to predict the fate of comparatively insoluble elements, such as plutonium.

So far no evidence has been found of currents strong enough to overcome gravity and bring claybound plutonium particles to the ocean surface. Most likely the material would remain on the seabed, unless it were carried up by creatures on the sea bottom. That prospect, and all other ways that radioactive materials might rise from deep-sea sediment layers to surface waters, warrant further investigation. The transportation of nuclear waste on the high seas also requires careful study. In particular, procedures would need to be developed for recovering lost cargo should a ship carrying radioactive materials sink or accidentally drop its load.

Engineers would probably seek to design the waste containers so that they could be readily retrieved from the bottom of the ocean in case of such a mishap or, in fact, even after their purposeful burial. Although subseabed disposal is intended to provide a permanent solution to the nuclear waste crisis, it may be necessary to recover material such as plutonium at some point in the future. That task would require the same type of drilling apparatus used for emplacement. With the location of the waste containers recorded at the time of interment, crews could readily guide the recovery equipment to the right spot (within a fraction of a meter) by relying on various navigation aids. At present, no non-nuclear nation has the deep-sea technology to accomplish this feat. In any event, performing such an operation in a clandestine way would be nearly impossible. Hence, the risk that a military or terrorist force could hijack the disposed wastes from under the seabed would be negligible.

Seafloor disposal would require a series of operations. After lowering a long, segmented drill pipe several kilometers to the ocean floor (a), technicians on the ship would put a "reentry cone" around the pipe and drop the device to the bottom (b). (The cone could guide another drill pipe to the hole later, should the need arise.) Turning and advancing the pipe (to which a bit is attached) would drill it into the ocean floor (c). By releasing the bit, the drillers could then lower a waste canister within the pipe using an internal cable (d). After packing that part of the hole with mud pumped down through the pipe (e), they would emplace other canisters above it (f). The topmost canister would reside at least some tens of meters below the seafloor (g). In about 1,000 years the metal sheathing would corrode, leaving the nuclear waste exposed to the muds (h). In 24,000 years (the radioactive half-life of plutonium 239), plutonium and other transuranic elements would migrate outward less than a meter (i).

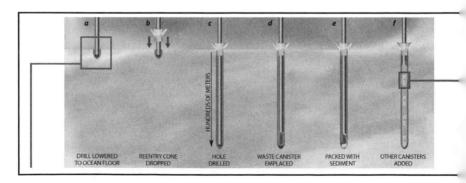

All Eggs in One Basket

The overall cost of a concerted program to evaluate sub-seabed disposal might reach $250 million—admittedly a large sum for an oceanographic research endeavor. But it is a relatively modest price to pay considering the immense benefits that could result. (As a point of comparison, about $2 billion has already been spent on site evaluation at Yucca Mountain, and another billion or two will probably be needed to complete further studies and secure regulatory approval. No actual construction, save for exploratory tunneling, has yet begun.) Yet no nation seems eager to invest in any research at all on subseabed disposal, despite the fact that it has never been seriously challenged on technical or scientific grounds. For example, a 1994 report by the National Academy of Sciences that reviewed disposal options for excess weapons plutonium called subseabed disposal "the leading alternative to mined geologic repositories" and judged implementation to be "potentially quick and moderate to low cost." But the academy panel stopped short of recommending the approach because of the anticipated difficulties in gaining public acceptance and possible conflicts with international law.

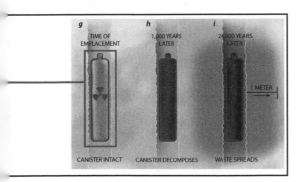

g TIME OF EMPLACEMENT   h 1,000 YEARS LATER   i 24,000 YEARS LATER   1 METER

CANISTER INTACT   CANISTER DECOMPOSES   WASTE SPREADS

Convincing people of the virtues of subseabed burial is, admittedly, a tough sell. But so is the Yucca Mountain project, which is strongly opposed by state officials and residents of Nevada. Subseabed disposal may turn out to be easier to defend among the citizenry than land-based nuclear waste repositories, which are invariably subject to the "not in my backyard" syndrome.

In any case, subseabed disposal is certain to evoke significant opposition in the future should the idea ever go from being a remote possibility to a serious contender. Oddly, the concept has recently come under direct fire, even though no research has been done in more than a decade. A bill introduced last year in the House of Representatives contains a provision that would prohibit the subseabed disposal of spent nuclear fuel or high-level radioactive waste as well as prevent federal funding for any activity relating to subseabed disposal—apparently including research. The intent of part of this bill is reasonable: subseabed disposal should be illegal until outstanding safety and environmental issues can be resolved. But it makes absolutely no sense to ban research on a technically promising concept for the disposal of weapons plutonium and high-level nuclear wastes.

Subseabed disposal faces serious international hurdles as well. In 1996, at a meeting sponsored by the International Maritime Organization, contracting parties to the so-called London Dumping Convention voted to classify the disposal of nuclear material below the seabed as "ocean dumping" and therefore prohibited by international law. This resolution still awaits ratification by the signatory nations, and the outcome may not be known for several years. But regardless of how that vote goes, we submit that "ocean dumping" is a wholly inappropriate label. It makes as much sense as calling the burial of nuclear wastes in Yucca Mountain "roadside littering."

Yet even assuming that the nations involved uphold the ban, the bylaws of the London convention would allow for subseabed disposal to be reviewed in 25 years, an interval that would provide sufficient time to complete a comprehensive appraisal of this disposal method. The 25-year moratorium could be wisely spent addressing the remaining scientific and engineering questions as well as gaining a firmer grasp of the economics of this approach, which remains one of the biggest uncertainties at present. In our most optimistic view, the legal infrastructure already established through the London convention could eventually support a program of subseabed disposal on an international basis.

A parallel effort should be devoted to public education and discussion. Right now there seems to be a strong aversion among some environmental advocates to any action at all to address the nuclear waste problem—and a solution that involves the oceans seems particularly unpalatable. But it makes no sense to dismiss the possibility of disposal in stable sub-oceanic formations—which exceed the land area available for mined repositories by several orders of magnitude—simply because some people object to the concept in general. It would be much more prudent to base a policy for the disposal of nuclear waste, whose environmental consequences might extend for hundreds of thousands of years, on sound scientific principles.

Barring a miraculous technical breakthrough that would allow radioactive elements to be easily transformed into stable ones or would provide the safe and economic dispatch of nuclear wastes to the sun, society must ultimately find somewhere on the planet to dispose of the by-products of the decades-long nuclear experiment. Americans in particular cannot responsibly pin all hopes on a single, undersized facility in a Nevada mountainside. They owe it to future generations to broaden their outlook and explore other possibilities, including those that involve the thick, muddy strata under the sea.

The Authors

Charles D. Hollister and Steven Nadis began regular discussions about subseabed disposal of nuclear wastes in 1995. Hollister, who is a vice president of the corporation of Woods Hole Oceanographic Institution, has studied deep-sea sediments for the past three decades. He continues to do research in the department of geology and geophysics at Woods Hole. Nadis graduated from Hampshire College in 1977 and promptly joined the staff of the Union of Concerned Scientists, where he conducted research on nuclear power, the arms race and renewable energy sources. He then wrote about transportation policy for the World Resources Institute. Currently a Knight Science Journalism Fellow at the Massachusetts Institute of Technology, Nadis specializes in writing about science and technology.

Current nuclear reactors produce energy by fission reactions. Fission occurs when the nucleus of an atom is split into two or more smaller pieces known as fission products. These products include two smaller nuclei, two or three free neutrons, and some protons. The process of splitting one nucleus into two or more smaller nuclei releases an enormous amount of energy, much more than in chemical reactions. It is this energy that is released and stored in nuclear fission reactors. Fission products are usually highly radioactive because the new, smaller nuclei are not stable isotopes. When the isotopes decay, they release gamma rays and radiation, becoming nuclear waste.

The opposite of fission is fusion, the combination of two atomic nuclei to make a larger nucleus. When the nuclei involved are lighter than iron, their fusion will release energy. The fusion of nuclei heavier than iron will absorb energy from the environment. Fusing nuclei requires *high temperatures and a lot of energy (though generally less than what will be generated by the fusion), which has thus far made it less practical than fission as an energy source. Because fusion creates much less radioactive waste than fission does, however, it is an attractive potential source of energy. Some scientists have claimed to achieve cold fusion (a means of fusing nuclei that requires much less energy and lower temperatures), but other laboratories have had difficulty replicating the results. Most researchers remain skeptical of the practical potential of cold fusion.*

The following article discusses work on hot fusion that continues today. The International Thermonuclear Experimental Reactor, a fusion reactor described here, received the go-ahead for construction in Cadarache, France, in June 2005. The magnetic fusion device will take ten years to build and is intended as an intermediate stage of development between small-scale fusion experiments and fusion-based commercial power plants. —LEH

"Fusion"
by Harold P. Furth
Scientific American, September 1995

During the 1930s, when scientists began to realize that the sun and other stars are powered by nuclear fusion, their thoughts turned toward re-creating the process, at first in the laboratory and ultimately on an industrial scale. Because fusion can use atoms present in ordinary water as a fuel, harnessing the process could assure future generations of adequate electric power. By the middle of the next century, our grandchildren may be enjoying the fruits of that vision.

The sun uses its strong gravity to compress nuclei to high densities. In addition, temperatures in the sun are extremely high, so that the positively charged nuclei have enough energy to overcome their mutual electrical repulsion and draw near enough to fuse. Such resources are not readily available on the earth. The particles that fuse most easily are the nuclei of deuterium (D, a hydrogen isotope carrying an extra neutron) and tritium (T, an isotope with two extra neutrons). Yet to fuse even D and T, scientists have to heat the hydrogen gases intensely and also confine them long enough that the particle density multiplied by the confinement time exceeds 10^{14} seconds per cubic centimeter. Fusion research since the 1950s has focused on two ways of achieving this number: inertial confinement and magnetic confinement.

The first strategy, inertial confinement, is to shine a symmetrical array of powerful laser beams onto a spherical capsule containing a D-T mixture. The radiation vaporizes the surface coating of the pellet, which explodes outward. To conserve momentum, the inner sphere of fuel simultaneously shoots inward. Although the fuel is compressed for only a brief moment—less than 10^{-10} second—extremely high densities of almost

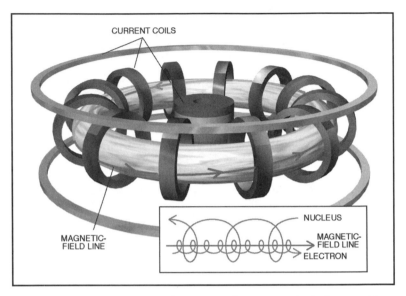

Tokamak Fusion Test Reactor (*above, top*), a magnetic fusion machine at Princeton, N.J., has achieved the highest energies to date. The hot fuel, consisting of deuterium and tritium nuclei, is confined by magnetic-field lines. These lines are generated by electric currents flowing around a doughnut-shaped container (*above, bottom*). The hot gas causes the inner walls, made of carbon, to glow pink.

10^{25} particles per cubic centimeter have been achieved at the Nova laser facility at Lawrence Livermore National Laboratory.

For more energetic lasers, the compression will be higher, and more fuel will burn. A future machine, the National Ignition Facility, the design and funding for which will be submitted to the U.S. Congress for final approval in 1996, will feature a 192-beam laser applying 1,800 kilojoules of energy within a few billionths of a second. If all goes well, fusion with this machine will liberate more energy than is used to initiate the capsule's implosion. France is planning to build a laser of similar capabilities near Bordeaux in 1996.

Magnetic Fields

The many magnetic fusion devices explored—among them stellarators, pinches and tokamaks—confine the hot ionized gas, or plasma, not by material walls but by magnetic fields. The most successful and highly developed of these devices is the tokamak, proposed in the early 1950s by Igor Y. Tamm and Andrei D. Sakharov of Moscow University. Electric current flows in coils that are arranged around a doughnut-shaped chamber. This current acts in concert with another one driven through the plasma to create a magnetic field spiraling around the torus. The charged nuclei and an accompanying swarm of electrons follow the magnetic-field lines. The device can confine the plasma at densities of about 10^{14} fuel particles per cubic centimeter for roughly a second.

But the gas also needs to be heated if it is to fuse. Some heat comes from the electrical resistance to the current flowing through the plasma. But more intense heating is required. One scheme being explored in tokamaks around the world uses radio-frequency systems similar to those in microwave ovens. Another common tack is to inject energetic beams of deuterium or tritium nuclei into the plasma. The beams help to keep the nuclei hotter than the electrons. Because it is the nuclei that fuse, the available heat is therefore used more efficiently. (The strategy represents a departure from earlier experiments that tried to imitate the sun in keeping the temperatures of the nuclei and the electrons roughly equal.)

Such a "hot ion" mode was used in 1994 at Princeton University's Tokamak Fusion Test Reactor (TFTR) to generate more than 10 million watts of fusion power. Although achieved for only half a second, the temperature, pressure and energy densities obtained are comparable with those needed for a commercial electrical plant. In 1996, during the next operational phase of the Joint European Torus (JET) at Culham, England, the experimenters may approach breakeven, generating as much energy as is fed into the plasma. At the JT-60U device at Naka, Japan, scientists are developing higher-energy injectors.

Apart from keeping the plasma hot, one challenge is to clean out continuously the contaminant atoms that are knocked off when nuclei happen to strike the walls. Several tokamaks have extra magnetic coils that allow the outer edges of the plasma to be diverted into a chamber where the impurities are extracted along with some heat. This system works well for present-day experiments, in which the plasma is confined for at most a few seconds. But it will not suffice for commercial power plants that will generate billions of watts during pulses that last hours or days. Researchers at JET and at the DIII-D tokamak in San Diego are attacking this problem.

Currently it is within our capability to construct and operate a tokamak that will sustain a stable, fusing plasma, not for fractions of a second but for thousands of seconds. The International Thermonuclear Experimental Reactor (ITER), a collaborative effort of the European Union, Japan, the Russian Federation and the U.S., aspires to do just that. ITER is expected to be a large machine, with a plasma about 16 meters in major diameter, featuring superconducting coils, tritium-breeding facilities and remote maintenance. The present schedule calls for a blueprint to be completed in 1998, whereupon the participating governments will decide whether to proceed with construction.

Looking Ahead

While ITER represents a valuable opportunity for international collaboration, there is some chance that its construction will be delayed. In the interim, experimenters at TFTR and other large tokamaks around the world will explore new ideas for making radically smaller and cheaper reactors, possibly influencing ITER's design. One of these proposals uses a complex twisting of the plasma to reduce heat loss greatly. Normally the magnetic field that winds around the torus has a higher twist near the center of the plasma. If instead the twist decreases in regions near the center, the plasma should be less turbulent, thus allowing higher pressures to be sustained. Experimentation, along with theoretical analysis and computer simulations, will vastly improve our ability to control such processes.

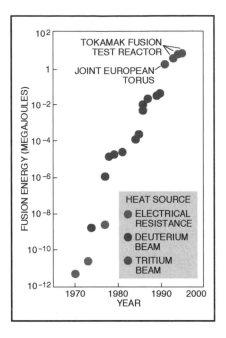

Energy release from fusion in tokamaks around the world has increased steadily in the past decades. Much of the improvement has come from heating the fuel by injection of energetic nuclei.

One more hint of how fusion could evolve in the next 25 years is in the utilization of the by-products. In a self-sustaining reaction the heat lost is made up by energy generated through fusion. Eighty percent of the fusion power is carried away by neutrons, which, being electrically neutral, slip through the confining magnetic fields. Trapped in outer walls, the neutrons give up their energy as heat, which is used to generate steam. The vapor in turn drives an engine to create electricity. The remaining 20 percent of the power goes to the other product of the fusion reaction, an alpha particle: a pair of protons and a pair of neutrons, bound into a nucleus. Being positively charged, alphas are trapped by the magnetic fields.

The alphas bounce around inside the plasma, heating up the electrons; the fuel nuclei are heated indirectly through collisions with electrons. Nathaniel J. Fisch of

Princeton and Jean-Marcel Rax of the University of Paris suggested in 1992 that rather than wasting valuable energy on the electrons, the alpha particles could help amplify special waves injected into the plasma that channel the energy directly to the nuclei. Thus concentrating the energy in the fuel could double the density of fusion power achieved.

The particles produced as by-products of fusion may be put to another, quite different, use. In this respect, taking a hint from history might be beneficial for fusion's short-term future. Two centuries ago in England, the industrial revolution came about because horses refused to enter coal mines: the first engines were put together to haul out coal, not to power cars or airplanes. John M. Dawson of the University of California at Los Angeles has proposed that during the next 20 to 30 years, while fusion programs are developing a technology for large-scale energy production, they could provide other benefits. For example, the protons formed as by-products of some fusion reactions may be converted to positrons, particles that can be used in medically valuable positron emission tomography scans.

During this phase of special applications, an abundance of new ideas in plasma physics would be explored, ultimately yielding a clear vision about future reactor design. Fifty years from now engineers should be able to construct the first industrial plants for fusion energy. Although far removed from immediate political realities, this schedule matches the critical timescale of

50 to 100 years in which fossil-energy resources will need to be replaced.

The Author

Harold P. Furth, professor of astrophysical sciences at Princeton University, directed the Princeton Plasma Physics Laboratory from 1980 to 1990. After earning a Ph.D. in high-energy physics from Harvard University in 1960, he moved to Lawrence Livermore Laboratory and then to Princeton to work on magnetic fusion. Furth is a member of the National Academy of Sciences and the American Academy of Arts and Sciences; he has received, among other honors, the E. O. Lawrence Memorial Award from the Atomic Energy Commission. He acknowledges the assistance of Kevin M. McGuire, also at the Princeton Plasma Physics Laboratory, in preparing this manuscript.

3 | Fuel Cells and Energy Storage

Fuel cells are devices that convert a fuel to electricity through chemical reactions. Electrons are removed from the fuel, travel through a circuit (generating the electricity), and then combine with oxygen to create water. The automobile industry is developing vehicles in which fuel cells take the place of the internal combustion engine as a potential environmentally friendly replacement for the gas-powered automobiles we know today.

Standard engines burn petroleum-based fuel. This combustion releases a variety of pollutants into the air, contributing to smog and global warming. The only by-product created by the use of fuel cells, on the other hand, is water. Automakers have focused on hydrogen as the fuel of choice for fuel-cell cars. Unfortunately, most hydrogen today is made from natural gas, a fossil fuel, so fuel cells are not yet as green as they might be.

Before fuel-cell cars become widely available, many issues must be addressed. First and foremost, an extensive network of hydrogen refueling stations is necessary because current fuel-cell vehicles cannot carry enough hydrogen for long trips. The relatively short distances that can be driven before refueling with hydrogen is another major and related problem. With many auto companies working on fuel-cell cars, however, these issues are sure to be addressed. The following article looks at the current state of fuel-cell car research. —LEH

"On the Road to Fuel-Cell Cars"
by Steven Ashley
Scientific American, March 2005

The automated speed traps that ward the approaches to Nabern, Germany, seem to be the only things that can wipe the smile off Rosario Berretta's face. "Please slow down here," he murmurs darkly as our vehicle nears the outskirts of the picturesque Swabian village. Berretta leads a team that is preparing a fleet of 60 of DaimlerChrysler's latest hydrogen fuel-cell car, the F-Cell, for testing worldwide. The aim is to allow automakers to evaluate the pollution-free, energy-efficient vehicles in diverse driving conditions. The curly-headed engineer is eager for visitors to try out the F-Cell's quick pickup off the line, one of the benefits of having an electric motor under the hood. But such maneuvers have to wait until the sharp eyes of the camera traps get small in the rearview mirror.

Despite its high-tech propulsion system, the F-Cell looks, performs and handles like a Toyota Corolla, a

Ford Focus or any other conventional small car. Thus, the F-Cell seems less like a next-generation prototype and more like a real-world car. The sole clue of anything out of the ordinary is the unfamiliar whir of a compressor—a noise that Berretta vows company engineers will soon muffle.

DaimlerChrysler is not alone in its quest for the ultimate clean vehicle. After a decade of focused research and development, the auto industry worldwide has passed a milestone with the arrival of the first test fleets of seemingly roadworthy fuel-cell cars. Twenty of Honda's latest FCX and 30 of Ford's Focus FCV fuel-cell powered compacts will soon be on the highways. General Motors plans to provide 13 fuel-cell-powered vehicles to the New York City metropolitan area for evaluation next year. Already 30 DaimlerChrysler-built fuel-cell buses are plying the streets of 10 European cities, and three more will shortly each be servicing Beijing and Perth.

Meanwhile nearly every other car company—particularly, Toyota but also Nissan, Renault, Volkswagen, Mitsubishi and Hyundai, among others—is operating at least a few prototype vehicles as well, one indication of the substantial funds carmakers are investing to perfect the technology. Today between 600 and 800 fuel-cell vehicles are reportedly under trial across the globe. And suppliers have emerged to develop and provide the components needed to build the prototypes. If all goes well, these developments will mark a midway milestone on the road to the initial commercialization of the fuel-cell car by the early part of the next decade.

Faced with ever tighter governmental regulatory limits on exhaust emissions, forecasts of impending oil shortages and a potential global warming catastrophe caused by greenhouse gases, the motor vehicle industry and national governments have invested tens of billions of dollars during the past 10 years to bring to reality a clean, efficient propulsion technology that is intended to replace the venerable internal-combustion (IC) engine [see "Vehicle of Change," by Lawrence D. Burns, J. Byron McCormick and Christopher E. Borroni-Bird; SCIENTIFIC AMERICAN, October 2002]. Critics, however, still question the industry's actual interest in producing a truly green car and whether this R&D effort is really enough to yield success anytime soon. Suspicions linger that work on fuel-cell vehicles is a smokescreen intended to shield business as usual long into the future. Car company executives reply that they foresee no better option to the hydrogen fuel-cell vehicle in the long run, because all alternatives, such as hybrid vehicles (which combine IC engines with electrochemical batteries), still burn petrochemical fuels and produce carbon dioxide and pollutants.

Stumbling Blocks

A two-hour drive, say, the 140 or so miles from Nabern to Frankfurt am Main on the German autobahn, would be enough to reveal the most telling distinction between the F-Cell and your typical IC engine car. In something less than 90 minutes, you would be stuck on the roadside out of fuel and with little prayer of finding a fill-up. Neither the F-Cell nor any of its hydrogen-powered kindred carries enough fuel to get anywhere near the 300-mile minimum driving range that car owners expect. And because hydrogen service stations are still few and far between, refueling would be problematic at best. So despite bright hopes and the upbeat pronouncements by automakers, considerable technical and market challenges remain that could delay introduction of the fuel-cell family car for years, if not decades.

Before early adopters can trade in their Toyota Priuses and Honda Accord Hybrids for something even greener, car manufacturers and their suppliers must somehow figure out how to do several things: boost onboard hydrogen storage capacity substantially, cut the price tags of fuel-cell drive trains to a hundredth of the current costs, increase the power plants' operating lifetimes fivefold, and enhance their energy output for SUVs and other heavy vehicles. Finally, to operate these vehicles, a hydrogen fueling infrastructure will be required to replace the international network of gas stations.

Even some of the automakers remain unconvinced that all this will happen soon: "High-volume production could be 25 years off," says Bill Reinert, national manager for Toyota's advanced technology group. "I'm less than hopeful about reducing costs sufficiently, and I'm quite pessimistic about solving hydrogen storage issues and packaging these large systems in a marketable vehicle." One telling sign that fuel-cell vehicles are still works in progress: nearly all car company representatives call for more government investment in basic research and hydrogen distribution systems to help overcome these roadblocks.

Stack Issues

A fuel-cell car, bus or truck is essentially an electric vehicle powered by a device that operates like a refuelable battery. Unlike a battery, though, a fuel cell does not store energy; it uses an electrochemical process to generate electricity and will run as long as hydrogen fuel and oxygen are fed to it.

At the core of the automotive fuel cell is a thin, fluorocarbon-based polymer—a proton-exchange membrane (PEM)—that serves as both the electrolyte (for charge transport) and a physical barrier to prevent mixing of the hydrogen fuel and the oxygen. Electricity for powering a fuel-cell car is produced when electrons are stripped from hydrogen atoms at catalysis sites on the membrane surface. The charge carriers—hydrogen ions or protons—then migrate through the membrane and combine with oxygen and an electron to form water, the only exhaust produced. Individual cells are assembled into what are called stacks.

Engineers chose PEM fuel cells because they convert up to 55 percent of the fuel energy put into them into work output; the efficiency figure for an IC engine is approximately 30 percent. Other benefits include relatively low-temperature operation (80 degrees Celsius); reasonably safe, quiet performance; easy operation; and low maintenance requirements.

The prospect of a commercial fuel-cell car by 2015 will depend on improvements in membrane technology, which makes up as much as 35 percent of the cost of a fuel-cell stack. Researchers list several needed enhancements such as low fuel crossover from one side of a membrane to the other, augmented chemical and mechanical stability of the membrane for greater durability, control over undesired by-reactions, and higher tolerance to contamination by fuel impurities or from unwanted reaction by-products such as carbon monoxide. Most of all, what is required is an across-the-board reduction in costs.

News of a "breakthrough" in membrane technology created a considerable stir in fuel-cell research circles last fall. PolyFuel, a small company in Mountain View, Calif., announced that it had created a hydrocarbon polymer membrane that it says offers superior performance and lower costs than current perfluorinated membranes. "It looks like a piece of sandwich wrap," James Balcom says, chuckling. The PolyFuel chief executive boasts a variety of reasons why his cellophanelike film performs better than the more common per-fluorinated membranes, notably DuPont's Nation material. The hydrocarbon membrane can run at higher temperatures than current membranes—up to 95 degrees C, which allows the use of smaller radiators to dissipate heat. It lasts 50 percent longer than fluorocarbon versions, he claims, while generating 10 to 15 percent more power and operating at lower (less troublesome) humidity levels. And whereas fluoro-carbon membranes cost about $300 per square meter, the PolyFuel materials potentially cost half the price. Although many other researchers remain skeptical about hydrocarbon membranes, Honda's newest FCX fuel-cell cars incorporate them.

Catalyst Conundrum

The other key to the operation of a PEM membrane is the thin layer of platinum-based catalyst that coats both of its sides and that represents 40 percent of the stack cost. The catalyst prepares hydrogen (from the fuel) and oxygen (from the air) to take part in an oxidation reaction by assisting both molecules to split, ionize, and release or accept protons and electrons. On the hydrogen side of the membrane, a hydrogen molecule (containing two hydrogen atoms) must attach to two adjacent catalyst sites, thereby freeing positive hydrogen ions (protons) to travel across the membrane. The complex reaction on the oxygen side occurs when a hydrogen ion and an electron mate with oxygen to produce water. This latter sequence must be finely controlled because it can yield destructive by-products such as hydrogen peroxide, which degrades fuel-cell components.

Because of the high cost of the precious metal ingredients, researchers are searching for ways to lower the platinum content. Their efforts include not only finding methods to raise the activity of the catalyst so less can be used for the same power output but also determining how to form a stable catalyst structure that does not degrade over time and avoiding side reactions that contaminate the membrane. One recent success in boosting catalytic activity was achieved by 3M Corporation researchers, who created nanotextured membrane surfaces covered with "forests of tiny columns" that significantly increased the catalysis area. Other work has concentrated on materials ranging from nonprecious metal catalysts such as cobalt and chromium to catalysts consisting of fine dispersions of particles embedded in porous composite structures.

Onboard Storage

One of the biggest worries among proponents of fuel-cell vehicles is how engineers will manage to stuff enough hydrogen onboard to provide the driving range that consumers demand. Five to seven kilograms will take a car up to 400 miles, but current fuel-cell proto-types hold from 2.5 to 3.5 kilograms. "Nobody really knows how to store twice that amount in a reasonable volume," says Dennis Campbell, chief executive of Ballard Power Systems in Vancouver, the dominant fuel-cell-stack maker.

Typically hydrogen is stored in pressure tanks as a highly compressed gas at ambient temperature. Many engineering teams are working on doubling the pressure capacity of today's 5,000-psi (pounds per square inch) composite pressure tanks. But twice the pressure does not increase the storage twofold. Liquid-hydrogen systems, which store the fuel at temperatures below −253 degrees C, have been tested successfully but suffer from significant drawbacks: about one third of the energy available from the fuel is needed to keep the temperature low enough to preserve the element in a liquid state. And despite bulky insulation, evaporation through seals robs these systems every day of about 5 percent of the total stored hydrogen.

Several alternative storage technologies are under development, but no surefire advances have occurred. "There's quite a good distance between what can be demonstrated in the lab and a fully engineered storage system that's affordable, longlasting and compact," says Lawrence Burns, vice president for research and development and planning at GM.

Probably the foremost candidates for a storage technology are metal hydride systems in which various metals and alloys hold hydrogen on their surfaces until heat releases it for use. "Think of a sponge for hydrogen," explains Robert Stempel, chairman of ECD Ovonic, a part of Texaco Ovonic Hydrogen Systems, the leader in this area. The hydrogen gas is fed into the storage tank under pressure and chemically bonds to the crystal lattice of the metal in question through a reaction that absorbs heat. The resulting compounds are called metal hydrides. Waste heat from the stack is used to reverse the reaction and release the fuel. In January, GM and Sandia National

Laboratories launched a four-year, $10-million program to develop metal hydride storage systems based on sodium aluminum hydride.

Because metal hydride storage systems tend to be heavy (about 300 kilograms), researchers at Delft University of Technology in the Netherlands have developed a way to store hydrogen in water ice—as a hydrogen hydrate, in which hydrogen is trapped in molecule-size cavities in ice. Water, of course, is significantly lighter than metal alloys. This approach is unexpected because hydrogen hydrates are notoriously difficult to make, typically requiring low temperatures and extremely high pressures, on the order of 36,000 psi. Working with scientists at the Colorado School of Mines, the Delft team came up with a "promoter" chemical—tetrahydrofuran—that stabilizes gas hydrates under much less extreme pressure conditions, only 1,450 psi. Theoretically, it should be possible to get about 120 liters (120 kilograms) of water to store about six kilograms of hydrogen.

Freezing Stacks

Several hundred people gathered behind the state capitol building in Albany, N.Y., to hear Governor George E. Pataki welcome the lease by New York State of a pair of Honda FCX hydrogen fuel-cell cars one cold, blustery late November morning in 2004. What made the event notable was the temperature of the air. All previous fuel-cell vehicle demonstration programs had been situated in warmer climes to ensure that the fuel-cell stacks would not freeze up. In previous designs, subzero temperatures could convert any liquid water into expanding ice crystals that can puncture membranes or rupture water lines. Early in the year Honda engineers demonstrated that their fuel-cell units could withstand winter conditions, an important engineering achievement for the fuel-cell research community.

After the speech, Ben Knight, vice president of R&D for American Honda, explained that the new freeze-resistant 2005 FCX models will start up repeatedly at –20 degrees C. Other car companies, including DaimlerChrysler and GM, have also claimed success with cold-starting test stacks in the lab.

Besides its ability to start up in midwinter temperatures, the 2005 version of Honda's FCX fuel-cell car—a four-seat compact hatchback—showcases other technical advances over the model released two years earlier. The new FCX is unusual, for example, because it employs an ultracapacitor—a device that stores energy in the electric fields between charged electrode plates—to provide short bursts of supplementary power for passing and hill climbing. Most other automakers use batteries for this purpose.

Infrastructure Issues

Later on that November day an even more enthusiastic crowd assembled for the second half of the planned ceremonies at the nearby headquarters of Plug Power, the Latham, N.Y.–based maker of stationary hydrogen fuel-cell energy units for backup power applications.

The cheering group of mostly Plug Power workers were there to celebrate the opening of a hydrogen fueling station that they had co-developed with Honda engineers. The Home Energy Station II contains a miniature chemical plant—a steam reformer—that extracts hydrogen fuel from piped-in natural gas using a steam-based process. "It's half the size of the previous version," said Roger Saillant, CEO of Plug Power. "Besides refueling vehicles, the system feeds some of the hydrogen into a fuel-cell stack to produce electricity for our headquarters building, which is also warmed in part by waste heat generated by the unit."

With great fanfare, one of the FCXs wheeled up to the fuel-dispensing pump—a metal box the size of a luxury kitchen stove that had been installed in the company parking lot. A state official first grounded the car by attaching a wire to the vehicle. He then dragged the fuel hose from the pump to the FCX's refueling port, inserted the nozzle and locked it into place. The unit finished filling the car's tank after about five or six minutes. Knight explained that the pump produces enough purified hydrogen to refill a single fuel-cell vehicle a day.

Afterward, Knight discussed the problems facing the development of a hydrogen infrastructure: "It's the classic chicken-and-egg dilemma," he said. "There's no demand for cars and trucks with limited fueling options, but no one wants to make the huge investment to create a fueling infrastructure unless there are fleets of vehicles on the road. So the question is: How do we create demand?" [see "Questions about a Hydrogen Economy," by Matthew L. Wald; SCIENTIFIC AMERICAN, May 2004].

A study by GM has estimated that $10 billion to $15 billion would pay to build 11,700 new fueling stations—enough so a driver would always be within two miles of a hydrogen station in major urban areas and so there would be a station every 25 miles along main highways. That concentration of mostly urban hydrogen stations would support an estimated one million fuel-cell vehicles, it says. "Twelve billion dollars, that's chump change when cable operators are plunking down $85 billion for cable system installations," exclaims Ballard's Campbell.

The Latham filling station—along with several dozen others scattered from Europe to California to Japan— embodies the first halting steps toward the construction of an infrastructure. Soon, Campbell says, about 70 hydrogen refueling stations will be operating worldwide, and California's Hydrogen Highway program has set a goal of 200 stations.

A National Academy of Sciences committee recently estimated that the transition to a "hydrogen economy" will probably take decades, because tough challenges remain. These include how to produce, store and distribute hydrogen in sufficient quantities and at reasonable cost without releasing greenhouse gases that contribute to atmospheric warming. Unfortunately, the extraction of hydrogen from methane generates carbon dioxide, a major greenhouse gas. If the energy sources for electrolysis (the splitting of water into hydrogen and oxygen using electricity) burn fossil fuels, they, too, would generate carbon dioxide. And hydrogen is a highly leak-prone gas that could escape from cars and production plants into the atmosphere, which could set off chemical reactions that generate greenhouse gases. Finally, using fossil fuels to make hydrogen takes more energy than that contained in the resulting hydrogen itself.

Researchers at the Idaho National Engineering and Environmental Laboratory and Cerametec in Salt Lake City have developed a way to electrolyze water and produce pure hydrogen with far less energy than other methods. The team's work points to the highest-known production rate of hydrogen by high-temperature electrolysis. Their new method involves running electricity through water that has been heated to about 1,000 degrees C. As the water molecules break up, a ceramic sieve separates the oxygen from the hydrogen. The resulting hydrogen has about half the energy value of the energy put into the process, which is better than competing processes.

Hydrogen proponents contend that arguments over infrastructure constitute a red herring. "U.S. industry currently produces 50 million to 60 million tons of hydrogen per year, so it's not like there's no expertise in handling hydrogen out there," Campbell notes. But automakers have a somewhat different perspective. "Fifty to 60 percent of the problems we have with our fuel cells arise from impurities in the hydrogen we buy from industry," complains Herbert Kohler, vice president of body and power-train research at DaimlerChrysler. "The chemical industry needs to do their homework."

Byron McCormick, GM's executive director of fuel-cell activities, likens investment in building a hydrogen infrastructure in the 21st century to the investment in railroads in the 19th century or to the creation of the interstate highway system in the 20th century: "There'll be a point relatively soon at which these kinds of how-do-you-get-it-funded decisions will be more important than the technology," he predicts.

Resolution of the myriad remaining technical and market issues will determine whether the transportation linchpin of the proposed hydrogen economy, the commercial fuel-cell vehicle, arrives in 10 years or 50.

Steven Ashley is a staff technology writer and editor.

Cars get much of the attention, but vehicles are far from the only devices that could eventually be powered by fuel cells. Such cells are also likely to begin appearing in portable electronic devices such as cell phones and laptop computers. Fuel cells have a higher energy density compared to rechargeable batteries of a similar size. As such, users will not need to recharge as often. Their

*high energy density also makes fuel cells good
for use in remote sensors (such as those used to
collect weather and climate data on satellites or in
remote, unpopulated regions of the world), which
cannot be regularly tended by maintenance
technicians.*

*Fuel cells can also run on a variety of fuels
other than hydrogen, some of them surprising.
For example, researchers are working on fuel
cells that can run on organic waste, such as
that collected on farms or from polluted water.
Such a fuel cell could eventually remove some
contaminants from water at a treatment plant
while providing power for other purification
procedures.*

*The following article looks at large-scale,
stationary fuel cells designed to provide power
to homes and other buildings. Although these
have yet to become widely adopted, perhaps
someday soon you will have a power plant in
your basement. —LEH*

"The Power Plant in Your Basement"
by Alan C. Lloyd
Scientific American, July 1999

As deregulation of the electric utility industry dissolves
the monopoly once held by most power generators,
one repercussion has been increasingly long distances
between some buyers and sellers of electricity.

Nevertheless, within a decade or two, some customers may find themselves living in a home whose electricity comes not from a generating plant tens, hundreds or even thousands of kilometers away but rather from a refrigerator-size power station right in their own basements or backyards. Moreover, not just homes but shops, small businesses, hotels, apartment buildings and possibly factories may all be powered in the same way: by fuel cells in the range of five to 500 kilowatts.

Companies and industrial research laboratories in Belgium, Canada, Denmark, Germany, Italy, Japan, Korea and the U.S. have aggressive fuel-cell development efforts under way, and at least a few are already selling the units. In fact, a subsidiary of United Technologies has been offering fuel cells of up to 200 kilowatts for almost a decade. They have sold about 170 units, many of which are used for generation of both heat and power at industrial facilities or for backup power. They are also increasingly being used at wastewater treatment plants and in "green" facilities, which showcase environmentally sensitive technologies and design.

At present, the high cost of fuel cells has limited their use to these and very few other specialized applications, made feasible for the most part by generous government subsidies. Electricity from fuel cells now costs $3,000 to $4,000 per kilowatt, as opposed to $500 to $1,000 per kilowatt for the ordinary gas-fired combustion turbine commonly used by utilities. Another drawback is limited lifetimes; so far no commercial fuel cell has been in operation for more than 10 years, and utilities expect to get at least 20 years of useful service life from their generating equipment.

On the other hand, fuel cells have several very desirable features: they operate relatively cleanly and silently, can use a variety of fuels, and are generally unaffected by storms and other calamities. Because of these advantages, some observers believe fuel cells can become viable for a reasonably large group of applications when per-kilowatt prices reach about $1,500.

Developers will have to achieve a number of design and manufacturing improvements before fuel cells attain even that level of price and performance. The incentives for them to do so, however, are great. As concern mounts about the harmful environmental effects of greenhouse gases from conventional power plants, increasing use of fuel cells is expected to help move industrial societies toward a "hydrogen economy." Electricity will come mainly from fuel cells and other hydrogen-based devices and from solar cells, windmills and other renewable sources—which will also electrolyze water to contribute hydrogen for the fuel cells. The shift to this hydrogen-based energy infrastructure will accelerate in coming decades, particularly as oil supplies begin dwindling.

Have Fuel, Will Energize

Fuel cells are not new; in fact, the basic concept is well over a century old. Like batteries, fuel cells come in a variety of different types. Also like batteries, they

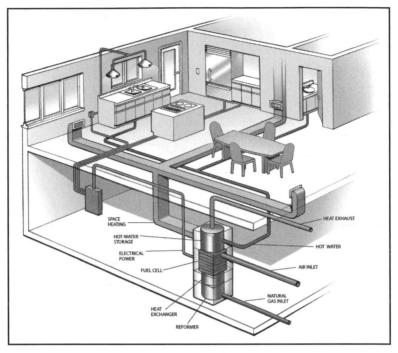

Solid-oxide fuel cell could provide electricity, heat and hot water to a home. The device operates at 800 degrees Celsius (1,500 degrees Fahrenheit), and some of the heat necessary to sustain such a temperature could be captured and directed into the home's heating ducts and into the hot-water tank. This use of heat that would otherwise be wasted enables the system to put as much as 90 percent of the fuel's chemical energy to productive use. Such a unit, which would produce up to 10 kilowatts of electricity, is being designed by Hydrogen Burner Technology in Long Beach, Calif., which plans to begin selling it around 2003.

produce an electric current by intercepting the electrons that flow from one reactant to the other in an electro-chemical reaction. A fuel cell consists of a positive and a negative electrode separated by an electrolyte, a material that allows the passage of charged atoms, called ions.

In operation, hydrogen is passed over the negative electrode, while oxygen is passed over the positive electrode. At the negative electrode, a highly conductive catalyst, such as platinum, strips an electron from each hydrogen atom, ionizing it. The hydrogen ion and the electron then take separate paths to the positive electrode: the hydrogen ion migrates through the electrolyte, while the electron travels on an external circuit. Along the way, these electrons can be used to power an electrical device, such as a lighting fixture or a motor. At the positive electrode, the hydrogen ions and electrons combine with oxygen to form water. (Interestingly, on the space shuttle, which gets its electricity from fuel cells, the by-product water is used for drinking.) To generate a useful amount of electric current, individual fuel cells are "stacked," like a club sandwich.

The device provides direct-current electricity as long as it is fed with hydrogen and oxygen. The oxygen typically comes from the ambient air, but the hydrogen usually comes from a system called a reformer, which produces the gas by breaking down a fossil fuel. One of the advantages of fuel cells is the great diversity of sources of suitable fuel: any hydrogen-rich material is a possible source of hydrogen. Candidates include ammonia, fossil fuels—natural gas, petroleum distillates, liquid propane and gasified coal—and renewable fuels, such as ethanol, methanol and biomass (essentially any kind of plant matter). Hydrogen can also be produced by solar, wind or geothermal plants. Even waste gases from landfills and water treatment plants will do. Reformers do release pollutants as they break down the fuel to make hydrogen. In comparison with a conventional gas-fired combustion turbine, however, the emissions are considerably less—typically a tenth to a thousandth, depending on the specific pollutant and how the emissions are controlled on the turbine.

Because a fuel cell's output is direct current, a device called an inverter is necessary to convert the DC into alternating current before the electricity can be of any practical residential or commercial use. In both the inverter and the reformer, power is lost, mostly as heat. Thus, although fuel cells themselves can have fuel-to-electricity efficiencies in excess of 45 percent, the energy losses in the reformer and inverter can bring the overall efficiency down to approximately 40 percent—about the same as a state-of-the-art gas-fired combustion turbine. As with the combustion turbine, however, recovering the waste heat—for example, to heat water or air—boosts the efficiency of the device significantly.

A popular misconception about fuel cells intended for stationary uses is that all of them are much more powerful than those being developed to propel automobiles. In fact, a fuel cell of just 40 to 50 kilowatts can easily meet the electrical needs of a large, four- or five-bedroom house or a small commercial establishment, such as a laundry. In comparison, because of the high level of power needed to accelerate a full-size car with four passengers, a fuel cell for a vehicle generally needs to put out at least 50 kilowatts. The more demanding requirements for automotive cells have prompted some observers to speculate that in the future some rural dwellers may even get power by simply plugging their homes into their cars.

The misconception about stationary fuel cells being uniformly larger than ones for automobiles probably stems from some very large experimental units tested by electric utilities over the past 20 years. The most notable of these were a 4.5-megawatt fuel cell installed by Consolidated Edison in New York City in 1982, a 4.5-megawatt unit operated by Tokyo Electric Power in 1984, an 11-megawatt unit operated by the same company from 1991 to 1997 and a two-megawatt plant tested by Pacific Gas and Electric in Santa Clara, Calif., in 1995. The U.S. demonstrations were rather problematic; the northern California cell, for example, rarely generated more than one megawatt—only half the capacity it was designed for—and the New York City cell never operated at all. The Japanese experiences, however, were much more favorable; the 11-megawatt unit, for example, ran for approximately 23,000 hours.

New Paradigm

Partly because of those difficulties, developers and proponents of stationary fuel cells have shifted to a paradigm based on a more decentralized approach.

Smaller units, of less than 50 kilowatts, will supply power to individual homes, and larger systems of up to several hundred kilowatts will power commercial buildings and other enterprises. Industry sources estimate that sales of the smaller fuel cells for residences and small businesses could reach $50 billion a year in the U.S. by 2030.

Such a figure may represent a certain amount of wishful thinking. There are no single-family residences at present that receive their power from onsite fuel cells; however, three companies—Plug Power in Latham, N.Y., Avista Laboratories in Spokane, Wash., and Northwest Power Systems in Bend, Ore.—have units that provide electricity to demonstration homes. The first fuel cell installed permanently at a home in the U.S., a brick, ranch-style house near Albany, N.Y., went into operation a little over a year ago, in June 1998.

Larger systems aimed at industrial or commercial uses are also in the works. At least one company hopes to introduce a 500-kilowatt fuel cell in the next few years for stationary applications, and several others are developing or selling fuel cells rated at 200 to 250 kilowatts. A 250-kilowatt cell could supply, for example, several stores in a strip mall or a small medical or corporate center.

Where larger loads are expected, multiple units can be linked. The developers of a building recently dedicated at 4 Times Square in New York City have installed two 200-kilowatt fuel cells to provide hot water for the building, light its facade and supply

Fuel Cells Compared

ELECTROLYTE	Proton-exchange membrane	Phosphoric acid	Molten carbonate	Solid-oxide ceramic
OPERATING TEMPERATURE	80° Celsius	Around 200° C	650° C	800–1,000° C
CHARGE CARRIER	Hydrogen ion	Hydrogen ion	Carbonate ion	Oxygen ion
REFORMER	External	External	Internal or external	Internal or external
PRIME CELL COMPONENTS	Carbon-based	Graphite-based	Stainless steel	Ceramic
CATALYST	Platinum	Platinum	Nickel	Perovskites (Titanate of calcium)
EFFICIENCY (percent)	40 to 50	40 to 50	Greater than 60	Greater than 60
STATUS OF DEVELOPMENT	Demonstration systems up to 50 kilowatts; 250-kilowatt units expected in next few years	Commercial systems operating, most of them 200-kilowatt; an 11-megawatt model has been tested	Demonstration systems up to 2 megawatts	Units up to 100 kilowatts have been demonstrated

backup power. The edifice is known as the Green Building because its developer, the Durst Organization, designed it partly to highlight technologies that are considered ecologically sound.

In a number of recent applications, fuel cells were chosen because their unusual features outweighed their high cost. For instance, a 200-kilowatt unit was installed at the police substation in New York City's Central Park. Use of a fuel cell obviated the expensive need to dig up the park to install underground power lines. In Nebraska the First National Bank of Omaha disclosed this past February that it would install four 200-kilowatt fuel cells at its Technology Center, where it processes credit-card transactions. The company chose fuel cells, backed up by auxiliary generators and by conventional electrical service from the local grid, because it needed extraordinarily high reliability for this application, in which even brief interruptions are quite costly.

Five Types of Cell

Of course, stationary fuel cells can be much larger and heavier than their mobile counterparts. So this market, though tiny today, has an unusual diversity of technologies being developed or sold. There are five main types of cell, each named after the electrolyte used in the system: phosphoric acid, molten carbonate, solid oxide, proton-exchange membrane and alkaline.

The phosphoric-acid fuel cell (PAFC) is the most mature technology of the five and the only one being offered commercially as of this writing in capacities above 100 kilowatts (all the fuel cells sold so far for commercial uses are PAFCs). Around the world, 12 organizations (seven in the U.S.) are marketing PAFCs or developing them. One of the largest is ONSI, a subsidiary of United Technologies, which has been deploying the units since the late 1980s. To date, the company has installed about 170 units, almost all of which are operating on natural gas. Some of these units have operated for tens of thousands of hours.

In the U.S., many of the purchases of ONSI fuel cells were subsidized under a program run by the Department of Defense and the Department of Energy since 1996. Buyers receive $1,000 per kilowatt or a third of the total project cost, whichever is lower. Already the program has distributed more than $18 million to the buyers and installers of more than 90 fuel-cell power plants.

A few large Japanese companies have sold approximately 120 PAFCs, with capacities ranging from 50 to

500 kilowatts. Several of these units have logged more than 40,000 hours of operational service.

In the U.S. and Japan, most PAFCs were purchased for power-generating installations that produce both heat and power. More recently, five other niche markets have sprung up, at landfills, wastewater treatment plants, food processors, power-generating facilities that cannot tolerate interruptions and green facilities (such as the aforementioned Green Building in New York City). In the first three applications, methane gas that would otherwise be an undesirable waste product is fed into the fuel cells; this free fuel helps to defray the cells' high purchase price.

The costs of PAFCs have been stalled for years now at about $4,000 per kilowatt—roughly three times what is believed to be necessary for the cells to be competitive. This fact has prompted some observers to write off the technology as a dead end, and most fuel-cell companies established in the past three or four years are pursuing other technologies, such as molten carbonate, solid oxide and proton-exchange membrane.

Molten-carbonate fuel cells (MCFCs) and solid-oxide fuel cells (SOFCs) are similar in that they must be operated at high temperatures, in excess of 650 degrees Celsius (1,200 degrees Fahrenheit). As its name implies, the MCFC cannot operate until its electrolyte becomes molten, and the solid-oxide cells rely on the high temperatures to reform fuels internally and ionize hydrogen, without a need for expensive catalysts. On the other hand, this heat comes from the cells' output, reducing it marginally. Some engineers are envisioning residential applications in which waste heat from these cells would be captured and used to heat living spaces and water [*see illustration on page 126*].

The major U.S. players in MCFCs are the Energy Research Corporation (ERC) in Danbury, Conn., and M-C Power Corporation in Burr Ridge, Ill. ERC built the two-megawatt plant in Santa Clara, Calif., mentioned earlier. It operated for 3,000 hours; unfortunately, it rarely put out more than a megawatt. Recently ERC changed its focus to 250-kilowatt units. M-C Power demonstrated a 250-kilowatt unit in San Diego in 1997, although it managed to produce only a disappointing 160 megawatt-hours of electricity before requiring repairs. Approximately 10 Japanese companies are also pursuing MCFCs.

As for the SOFC, a total of 40 companies around the world are developing the technology. One of the largest was created in 1998 when Siemens acquired Westinghouse Power Generation; both companies had been working on versions of the SOFC. Other important U.S. developers of the SOFC include SOFCo, ZTek Corporation and McDermott.

Proton Exchange and the New Paradigm

If the phosphoric-acid, molten-carbonate and solid-oxide fuel cells are all in some ways vestiges of the centralized paradigm of deployment, the proton-exchange membrane technology represents the burgeoning decentralized approach. There is mounting enthusiasm for PEM cells in the wake of recent, significant reductions in the cost of producing their electrolytes and of the creation of catalysts that are more resistant to the degradation caused by carbon monoxide from the reformers.

The key component of a PEM is a thin, semiper-meable membrane, which functions as an electrolyte. Positively charged particles—such as hydrogen ions—can pass through this membrane, whereas electrons and atoms cannot. A few years ago developers discovered that Gore-Tex, a material often used in outer garments, can be used to strengthen the membranes and signifi-cantly improve their operating characteristics.

That and other advances have prompted a flurry of activity in the devices. At present, some 85 organi-zations, including 48 in the U.S., are doing research on or developing PEMs. For example, Ballard Generation Systems in Burnaby, B.C., is working on a modular-design PEM that can be configured up to 250 kilowatts. The company hopes to begin selling the 250-kilowatt units in 2001.

General Electric's Power Systems division and the previously mentioned Plug Power have joined together to market, install and service PEMs worldwide with capacities up to 35 kilowatts. The joint venture expects to begin field-testing prototype units later this year and to install the first residential-size units early in 2001. Plug Power installed and operates a seven-kilowatt PEM fuel cell at a home in Latham, N.Y. (where two of the company's engineers live during the week).

Another company betting on PEMs is H Power Corporation in Belleville, N.J., which offers small units ranging from 35 to 500 watts. Besides promoting the cells for the usual residential uses, H Power has ventures targeting applications in backup power, telecommunications and transportation. In an unusual marketing strategy, it is even promoting fuel cells as security against the blackouts that some people fear will result from software glitches at the turn of the millennium. In another project, H Power is retro-fitting 65 movable message road signs with fuel-cell power sources for the New Jersey Department of Transportation.

A few of the other PEM developers are Avista Laboratories in Spokane, Wash., which is working in conjunction with the engineering firm Black and Veatch; Matsushita Electric Industrial Company in Japan, which is focusing on a 1.5- to 3.0-kilowatt cell; and Sanyo, which has developed an appliance-like one-kilowatt PEM fuel-cell system that operates on compressed hydrogen. Sanyo also plans to develop a two-kilowatt unit using either a natural gas or methanol reformer.

Alkaline fuel cells, which have a relatively long history in exotic uses such as the space shuttle, are intriguing because they have efficiencies as high as 70 percent. So far their very high cost and other concerns have kept them out of mainstream applications. Nevertheless, a few organizations are attempting to produce alkaline units that are cost-competitive with

other types of fuel cells, if not with other generation
technologies.

Providing Premium Power

Other than continued subsidies, the best hope for fuel
cells in the near future will be applications in which
electricity is already expensive or in which waste gas
can be used to fuel them. In fact, at current prices, it
will probably take a combination of subsidies and
unusually favorable circumstances. For instance, under
an expanded federal government initiative in the U.S.,
purchasers of residential-size fuel-cell power plants may
be eligible for federal help. In the past, federal agencies
provided assistance only for the purchase of units of
100 kilowatts or more.

In the more distant future, concerns about global
climate and related pressures to reduce emissions of
carbon dioxide might even pave the way for large-scale
use of fuel cells in the developing world. In a paper
presented last year, Robert H. Williams of the Princeton
University Center for Energy and Environmental
Studies proposed that fuel cells might play a role in the
electrification of China, whose 1.2 billion people have
one of the world's lowest per capita rates of electricity
usage. China has vast reserves of coal, which, Williams
noted, could be turned into a supply of hydrogen-rich
gas well suited to fuel cells. The challenge would be
to "decarbonize" the coal during gasification. This
decarbonization would produce waste carbon dioxide—
a key greenhouse gas. So engineers and geologists would

have to sequester it somehow, separating it permanently from the environment.

Because of such issues, any large-scale deployment of fuel cells in a developing country could be a decade off. But in the developed world over the next several years, improvements in the proton-exchange, molten-carbonate and solid-oxide technologies will enable fuel cells to carve out new niches and expand the small ones they already occupy. As they do so, they will begin ushering in a cleaner, more environmentally benign hydrogen economy—and perhaps not a moment too soon.

The Author

Alan C. Lloyd is chairman of California's Air Resources Board, part of the state's Environmental Protection Agency. When this article was commissioned, he was executive director of the Energy and Environmental Engineering Center at the Desert Research Institute in Nevada. Before that, he was chief scientist at the South Coast Air Quality Management District in California. He wishes to thank the U.S. Fuel Cell Council and Fuel Cells 2000 for their help in the preparation of this article.

"Net metering" allows homeowners and businesses that produce their own energy, such as from solar cells, to store the excess energy they

*don't immediately need in the power grid. They
can then withdraw power from the grid when
necessary. According to the U.S. Department of
Energy, thirty-five states had programs in place
in 2004 to provide net metering, more than twice
the number in 1997, when the following article
was written. The programs vary: Some are only
available to residential customers and not
businesses, for example. Others set limits on the
permissible types of energy generation; only
wind and solar power are eligible in California,
for instance.*

*Overall, nearly 7,000 customers in the United
States participated in net metering programs in
2003. That may seem small, but with solar panels
and other energy generation devices suitable for
home and business use becoming more widely
available and affordable, net metering programs
are likely to become more popular in coming
years. —LEH*

"Power to the People"
by David Schneider
Scientific American, May 1997

After an environmentally conscious home owner installs
solar panels on the roof or a wind-driven generator in
the backyard, clean electricity flows for free—but
only while the sun shines and the wind blows. One
way to cope with this intermittence is to store energy

in batteries. But a growing number of utility companies now allow home owners a less expensive option. They can deposit the excess electricity they produce into the power grid and withdraw it at later times using just their standard household electrical meter, which can run equally well backward or forward. Permitting such "net metering" gives a single home owner a privilege normally exercised only among giant utility companies: trading electricity generated at one time for the power required at another.

Net metering should help spur small-scale production of renewable energy. Otherwise, a home owner receives only the so-called avoided cost for any electricity exported to the power grid, and this rate is just a fraction of what the utilities typically charge residential customers. But with net metering, individuals can get, in essence, the full retail price for the electricity they generate, so long as they buy it back during the same billing period. (Any surplus production at the end of the month still earns only the wholesale price.)

Although net metering alone does not normally make home generation of renewable energy economical, it does bring such efforts somewhat closer to the break-even point. In sunny Hawaii, for instance—where the state government offers a solar-energy tax credit and the cost of electricity is especially high—net metering could make home solar-power systems cost-effective.

Net metering may also encourage people in more marginal situations to invest in solar or wind generators.

According to Michael L. S. Bergey, president of Bergey Windpower, when net metering is not offered to them, some home owners will balk at the idea of selling their excess electricity at a discount to the local utility, which then resells the power for several times the price. "The biggest thing [net metering] does is change the mind-set," concurs Christopher Freitas, director of engineering for Trace Engineering, a company that makes power conditioning equipment used in home installations. "The idea of being able to spin a utility meter backward really appeals to people."

Yet some utilities and government officials have resisted net metering, which is now available in Japan and Germany but only in 16 U.S. states. When advocates of renewable energy proposed a net-metering law in California in 1995, Pacific Gas and Electric, a major utility, fought against it—but lost. New York State governor George Pataki vetoed a net-metering law passed last year, citing concerns that energy from homes might continue to flow through lines during general outages, endangering power company workers.

"That argument was absolutely bogus," says Thomas J. Starrs, a lawyer at Kelso, Starrs and Associates, who helped to write the legislation for net metering in California. Power from home generators, he notes, runs through a device called an inverter that converts direct current to alternating current; should power in an area fail, the inverter automatically cuts off the flow. But green advocates recognize that they will still have to expend some energy of their own to persuade all concerned parties—from state governors and utility officials to local building inspectors and insurers—that individual home owners can safely generate power from sources that do not create radioactive waste or greenhouse gases. As Starrs quips, "We're still working out the kinks."

4 Alternative Energy Sources

Most people know methane as a smelly gas that is a major contributor to the distinctive odor of cattle ranches and fertilizer. It is also a major component of natural gas, which fuels millions of homes in the United States. With oil prices on the rise, the United States is increasingly relying on less expensive natural gas to satisfy energy demands. The Energy Information Administration predicts that U.S. consumption of natural gas will increase by more than 40 percent by 2025.

The following article explores a potential new source of natural gas: methane hydrate. A hydrate is a solid material that includes water in a specific ratio to the other components. Thus, methane hydrate is essentially frozen natural gas. First discovered when natural gas pipelines were extended into colder climates, methane hydrate was considered a nuisance for many years. Now it is regarded as a valuable resource.

Large deposits of methane hydrate have been discovered under the sea floor and under permafrost (permanently frozen soil) in Canada, Russia, and Alaska. Research is under way to learn more about this compound and determine the best way to collect and transport it. —LEH

"Flammable Ice"
by Erwin Suess, Gerhard Bohrmann, Jens Greinert, and Erwin Lausch
Scientific American, November 1999

It was a thrilling moment when the enormous seafloor sampler opened its metallic jaws and dumped its catch onto the deck of our ship, *Sonne.* A white substance resembling effervescent snow gleamed amid the dark mud hauled up from the bottom of the North Pacific Ocean. Watching it melt before our eyes, we sensed that we had struck our own kind of gold.

As members of the Research Center for Marine Geosciences (GEOMAR) at Christian Albrechts University in Kiel, Germany, we and our colleagues were searching for methane hydrate—a white, icelike compound made up of molecules of methane gas trapped inside cages of frozen water. To that end, we had undertaken several expeditions to inspect, with the help of a video camera tethered to the ship, a submarine ridge about 100 kilometers (62 miles) off the coast of Oregon. Earlier seismic investigations and drilling had suggested that this area might hold a substantial stash of our treasure. On July 12, 1996, we noticed peculiar white spots in the mud 785 meters (2,575 feet) below our ship.

To make sure this telltale sign of hydrate was the real thing, we directed our sampler, a contraption like a backhoe with two scoops, to take a giant bite out of the seafloor. Even while retrieving the payload, we saw our expectations confirmed. As the sampler ascended, the video camera mounted inside its jaws revealed that bubbles—attesting to the rapid escape of methane gas—were beginning to emerge from the muddy heap. Stable only at near-freezing temperatures and under the high-pressure conditions generated by the weight of at least 500 meters of overlying water, methane hydrates decompose rapidly above that depth. As the sample approached the ocean's surface, the flow of bubbles gradually increased and burst through the water's sparkling surface long before the jaws of the sampler did.

We wondered how much intact hydrate would reach the deck. Moving quickly, we managed to safeguard roughly 100 pounds (45 kilograms) of the hissing chunks in containers cooled with liquid nitrogen. In the end, we even had a few pieces left over, which inspired an impromptu fireworks display. Just holding a burning match to one of the white lumps ignited the hydrate's flammable methane, which also is one of the primary hydrocarbon components of natural gas. The lump burned with a reddish flame and left only a puddle of water as evidence of its former glory.

Before 1970 no one even knew that methane hydrates existed under the sea, and our haul was by far the largest quantity ever recovered from the ocean depths. Yet hydrates are by no means a rarity. On the contrary, in recent years they have been found to occur worldwide—from Japan to New Jersey and from Oregon to Costa Rica—in enormous quantities. Estimates vary widely, but most experts agree that marine gas hydrates collectively harbor twice as much carbon as do all known natural gas, crude oil and coal deposits on the earth [*see illustration on page 147*].

The energy stored in methane hydrates could potentially fuel our energy-hungry world in the future (if practical mining techniques are devised). But the hydrates also have a worrisome aspect: methane escaping from disturbed undersea hydrates may be an ecological threat. If even a small portion of these deposits decompose through natural processes, astonishing quantities of methane will be set free to exacerbate the greenhouse effect and global warming.

Although methane remains in the atmosphere relatively briefly—10 years on average—it does not vanish without a trace. In the presence of free oxygen, a methane molecule's single atom of carbon disengages from its four hydrogen atoms to become carbon dioxide, the most infamous of all greenhouse gases because it is one of those spewed into the atmosphere during the combustion of fossil fuels.

But are decomposing methane hydrates contributing to global warming now? And are they likely to do so in the future? Our 1996 journey—along with dozens of voyages and experiments since—have revealed something about the structure and origins of a variety of these massive yet remarkably unstable deposits and have provided some answers to the climate questions, but our understanding is far from complete. We and our

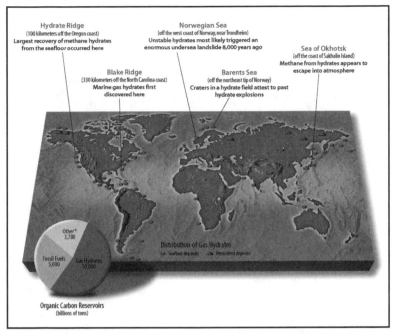

Hydrate deposits containing methane and other gases exist worldwide under the seafloor and in permafrost regions on land (*map*). Gas hydrates contain more organic carbon than does any other global reservoir (*pie chart*).

colleagues at GEOMAR continue our quest to understand just what role methane hydrates play in oceanfloor stability and both past and future climatic change.

Turning Heads

It was the immense cache of energy trapped in marine methane hydrates that first turned the heads of politicians. But the challenges of tapping that resource are now making some officials look the other way. Hydrates tend to form along the lower margins of continental slopes, where the seabed drops from the relatively shallow shelf, usually about 150 meters below the surface, toward the ocean's abyss several kilometers deep. The hydrate deposits may reach beneath the ocean floor another few hundred meters—deeper than most drilling rigs can safely operate. Moreover, the roughly sloping seafloor makes it difficult to run a pipeline from such deposits to shore.

Countries that wish to rely less on foreign fossil fuels have started to overcome these technical difficulties, however. Japan was scheduled to launch an experimental hydrate drilling project off the coast of Hokkaido in October. U.S. engineers also are playing with ideas for tapping hydrate energy sources. But as long as relatively cheap gas and oil remain available, most industrial countries are unlikely to invest heavily in the technologies needed to harvest hydrates efficiently.

Methane hydrates captured the attention of petroleum geologists a bit earlier than that of politicians. When engineers first realized in the 1930s that gas-laden ice crystals were plugging their gas and oil pipelines, laboratory researchers spent time studying hydrate structure and composition. For example, they learned that one type of hydrate structure consists of icy cages that can absorb small gas molecules such as methane, carbon dioxide and hydrogen sulfide. A different type forms larger cavities that can enclose several small molecules or larger hydrocarbon molecules, such as pentane. What is more, the individual cages can differ in the kinds of gas molecules they capture.

In the 1960s scientists discovered that hydrates could also form in natural environments. They found the first natural deposits in the permafrost regions of Siberia and North America, where the substances were known as marsh gas. In the 1970s geophysicists George Bryan and John Ewing of Lamont-Doherty Earth Observatory of Columbia University found the earliest indication that methane hydrates also lurk beneath the seafloor. The hint came from seismological studies at Blake Ridge, a 100-kilometer-long feature off the North Carolina coast.

Seismologists can distinguish layers beneath the seafloor because sound waves bounce off certain kinds of dirt and rock differently than off other kinds. Some 600 meters below the ocean floor Bryan and Ewing saw an unusual reflection that mimicked the contour of the ridge. Their conclusion: this bottom-simulating reflector was the boundary between a methane hydrate layer and a layer of free methane gas that had accumulated below. Other experts found similar features elsewhere, and soon this type of reflector was being mapped as a methane hydrate deposit in ocean basins around the globe.

We used a bottom-simulating reflector, along with underwater video cameras, to guide our 1996 search for methane hydrates along the North Pacific seafloor promontory that has since been named Hydrate Ridge. Our successful recovery of intact hydrate on that expedition made it possible to study this unusual material in detail for the first time. Being able to analyze the texture and chemistry of its microscopic structure allowed us to confirm the plausible but previously unproved notion that the methane derives from the microbial decomposition of organic matter in the sediment.

Most telling were chemical tests that showed the hydrates to be enriched in carbon 12. Inorganic methane that seeps out of volcanic ridges and vents has higher levels of carbon 13, an isotope of carbon with an additional neutron. But the bacteria that digest organic matter in oxygen-deficient conditions such as those in sediments at the bottom of the sea tend to sequester more carbon 12 in the methane they generate.

The seafloor off the coast of Oregon also proved an especially fruitful theater of operations for assessing the stability of methane hydrate deposits and their potential role in releasing carbon into the atmosphere. Combined with research from other sites, these analyses indicate that methane hydrate deposits can be disturbingly labile.

We now know that in places along Hydrate Ridge, the seabed is virtually paved with hundreds of square meters of hydrate. These deposits form part of the packet of sediments riding piggyback on the Juan de Fuca tectonic plate, which is sliding underneath North America at a rate of 4.5 centimeters (1.7 inches) a year. As the Juan de Fuca plate is subducted, the sediments and hydrates it carries are partially sheared off by the

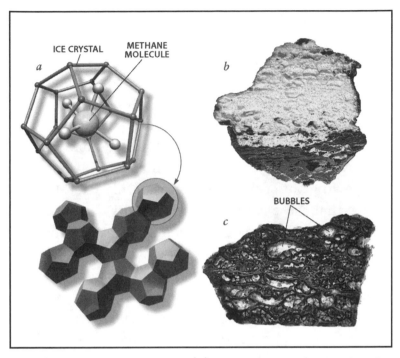

Crystalline cages of frozen water (*a*) sometimes snare molecules of methane gas that have been given off as microbes digest organic matter in the seafloor mud. The fresh hydrate sample shown above (*b*) formed a few meters beneath the seafloor off the coast of Oregon, where rising bubbles of methane gas were trapped underneath denser layers of mud. The methane reacted with the near-freezing water to form hydrates. Lens-shaped bubbles are clearly visible in another slice of the same methane hydrate sample (c), which was polished with a cold saw.

upper plate and pressed into folds or piled several layers high like a giant stack of pancakes. This distorted material forms a wedge of mud that accumulates against the North American plate in the shape of ridges running nearly parallel to the coast.

Methane Plumes

In 1984 one of us (Suess) was the first to observe
these ridges and their world of cold, eternal night
from *Alvin*, the research submersible operated by the
Woods Hole Oceanographic Institution. Outside *Alvin*'s
porthole, Suess saw a landscape of stone chimneys
built by minerals precipitated from hazes of water and
gas spewing out of the earth's crust. Only later did we
realize that these chimneys are partially the product
of a methane hydrate deposit being squeezed as one
tectonic plate scrapes past the other.

Plumes of gas and fluids also escape along faults
that cut through the sediment and gas hydrates alike.
Although these plumes, also called cold vents, are
unlike the hot springs that form along mid-oceanic
ridges (where hot lava billows out of a crack in the
seafloor), they are nonetheless warm enough to further
destabilize the hydrates. These melt when the surrounding temperature creeps even a few degrees above
freezing. As new hydrates form near the seafloor, the
lower portions melt away, with the result that the
overall layer migrates upward over time.

Melting at the bottom of a hydrate layer liberates
not only freshwater but also methane and small amounts
of hydrogen sulfide and ammonia. Oxidation of these
chemicals into carbon dioxide, sulfate and nitrate
provides nourishment to rich communities of chemical-
eating bacteria. These microbes in turn serve as food
for such creatures as clams and colonies of tube worms.

Such oases of life stand out on the otherwise sparsely inhabited seabed.

Our investigations also revealed that the gases liberated at these densely populated vents give rise to an immensely active turnover of carbon. Oxidation of the liberated methane generates bicarbonate, which combines with calcium ions in the seawater to form calcium carbonate, better known as limestone. Such limestone is what Suess saw in 1984 in the form of chimneys and vent linings along the crest of Hydrate Ridge—and what we now realize is a hint of a deeper hydrate layer.

Along the western flank of Hydrate Ridge, massive limestone blocks cover the crack created by a large fault. But despite the limestone casing and the activities of the vent organisms, surprising quantities of methane escape into the surrounding ocean water. In fact, we measured concentrations that are roughly 1,300 times the methane content of water at equilibrium with the methane content of the air. We still do not know how much of the methane is oxidized in the water and how much actually enters the air, but it is easy to imagine that an earthquake or other dramatic tectonic event could release large amounts of this highly potent greenhouse gas into the atmosphere.

Researchers at GEOMAR have a much better idea of how much methane escapes in plumes rising up from hydrate fields in the Sea of Okhotsk off the east coast of Asia. About as big as the North Sea and the Baltic Sea combined, this body of water is delineated by the Kamchatka Peninsula and the Kuril island arc. In the summer of 1998 a joint German-Russian team using fish-finder sonar documented methane plumes as tall as 500 meters billowing out of methane hydrate deposits on the seafloor. With our video camera tethered to the boat, we also saw giant chimneys reminiscent of the cold vents along Hydrate Ridge in the Pacific.

Even before we had visual evidence of the hydrate deposits, we knew that enormous quantities of methane accumulate under the blanket of ice that typically covers much of the Sea of Okhotsk for seven months a year. We measured a concentration of 6.5 milliliters of methane per liter of water just beneath the ice during a 1991 expedition. When the sea was free of ice the next summer, this figure was only 0.13—the difference had vented into the atmosphere. No similar methane flux has yet been observed anywhere else in the world, so this event may be unique. Still, this onetime measurement of methane escape from the Sea of Okhotsk clearly demonstrates that methane hydrates below the oceans can be a significant source of atmospheric methane. To help evaluate the possible current and future climatic impact of the methane, scientists are now sampling methane concentrations in the Okhotsk seawater every two months.

Shaky Ground

Plumes caused by seafloor faulting and the natural decomposition of hydrates can release methane slowly to the atmosphere, but it turns out that this process is sometimes much more explosive. In the summer of 1998 Russian researchers from the Shirshov Institute of Oceanology in Moscow found unstable hydrate fields off the west coast of Norway that they suspect are the cause of one of history's most impressive releases of trapped methane, an event known as the Storrega submarine landslide.

From previous explorations of the seafloor, scientists know that 8,000 years ago some 5,600 cubic kilometers (1,343 cubic miles) of sediments slid a distance of 800 kilometers from the upper edge of the continental slope into the basin of the Norwegian Sea, roughly at the latitude of Trondheim. The consequence of so much mud pushing water out of its path would have been devastating tsunamis—horrific swells that suddenly engulf the coastline.

The presence of methane hydrate fields in the same seafloor vicinity implies that unstable hydrates triggered the slide as they rapidly decomposed because of a change in the pressure or temperature after the last ice age. As the glaciers receded, the seafloor no longer had to support the enormous weight of the ice. As the land rebounded, the overlying sea and ice both warmed and became more shallow, suddenly moving the hydrates out of their zone of stability.

Could a geologic event like this strike again? Off the coast of southern Norway the risk of new slides appears to be relatively small, because the hydrate fields have for the most part decomposed. But the issue of the stability of the continental slope is assuming a

From the Seafloor to the Sky
How Methane Hydrates May Layer the Climate

Hydrate deposits may contribute to global warming by gradually releasing methane, a greenhouse gas, into the atmosphere (1–5). Catastrophic releases of methane from hydrates can warm the planet as well. The hydrate initially forms when gas bubbles rising through the seafloor sediments become trapped under denser layers of mud above. In this high-pressure environment, accumulated methane reacts with near-freezing water to form hydrates in the tiny spaces in the sediments.

—*The Editors*

1. Heat rising from the earth's interior melts the bottom of the hydrate layer and releases methane along faults or through the sediment above.
2. Methane escaping from the hydrate layer nourishes rich communities of microbes that in turn sustain cold-vent creatures such as tube worms.
3. Much of the methane oxidized by the bacteria combines with calcium ions in the seawater to form limestone crusts on the seafloor.
4. Some of the remaining methane rises in plumes to the ocean surface and escapes into the atmosphere.
5. In the atmosphere the methane converts to another greenhouse gas, carbon dioxide. Both

gases can gather into an insulating layer that
heats the lower atmosphere and thereby changes
climate patterns.

heightened importance in view of current global
warming and the strong possibility of further changes
in the earth's climate in the near future. Beyond contributing to tsunamis, hydrate formations that become
unstable and decompose will release methane into the
oceans. In fact, melting a mere cubic meter of hydrate
releases up to 164 cubic meters of methane, some of
which would surely reach the atmosphere. In turn, a
warming of the lower atmosphere would heat the
oceans, launching a vicious circle of more dissolution
of hydrates and more atmospheric warming.

Many researchers think that an explosive methane
release from a single large site can create dramatic
climate changes on short timescales. James P. Kennett,
an oceanographer at the University of California at
Santa Barbara, has hypothesized that catastrophic
releases of methane could have triggered the notable
increase in temperature that occurred over just a few
decades during the earth's last ice age some 15,000 years
ago. An international team led by former GEOMAR
member Jurgen Mienert, now at Tromso University
in Norway, recently found possible evidence of this
methane release on the floor of the Barents Sea, just
off Norway's northeastern tip.

There, fields of giant depressions reminiscent of
bomb craters pockmark the immediate vicinity of
methane hydrate deposits. Mienert's team measured
the biggest of these craters at 700 meters wide and
30 meters deep—a size clearly suggestive of catastrophic
explosions of methane. Whether these eruptions
occurred more or less simultaneously has not yet been

determined, but faults and other structural evidence indicate they probably took place toward the end of the last ice age, as Kennett proposed. The explosions very likely followed a scenario like the one suggested for the cause of the Storrega landslide: warming seas rendered the hydrates unstable, and at a critical point they erupted like a volcano.

Older Hints

Researchers have also uncovered evidence that methane liberated from gas hydrates affected the global climate in the more distant past—at the end of the Paleocene, about 55 million years ago. Fossil evidence suggests that land and sea temperatures rose sharply during this period. Many species of single-celled organisms dwelling in the seafloor sediment became extinct. A flux of some greenhouse gas into the atmosphere presumably warmed the planet, but what was the source? Carbon isotopes turned out to be the key to interpreting the cause of the rapid rise in temperature.

Scientists found a striking increase in the lighter carbon 12 isotope in the preserved shells of microscopic creatures that survived the heat spell. Methane hydrates are the likeliest source for the light carbon—and for the greenhouse gas flux—because these deposits are the only places where organic methane accumulates to levels that could influence the isotopic signature of the seawater when they melt, The carbon 12 enrichment characteristic of the hydrates disperses into the seawater with the liberated methane and persists in its oxidation product, carbon dioxide. Some of the carbon dioxide in turn becomes incorporated into the calcium carbonate shells of the sea creatures, while some of the methane makes its way to the atmosphere to help warm things up.

Gerald Dickens, now at James Cook University in Australia, used a computer simulation to test whether melting methane hydrates could have belched out enough greenhouse gases to subject the earth to a heat shock 55 million years ago. He and his former collaborators at the University of Michigan based their simulation on the assumption that hydrates corresponding to about 8 percent of today's global reserves decomposed at that time. Because liberated methane is converted immediately (on a geologic timescale) into carbon dioxide, they tracked this compound only.

In 10,000 simulated years, 160 cubic kilometers of carbon dioxide containing carbon 12 showed up in their model atmosphere every year. Adding this carbon dioxide caused the lower atmosphere to warm by two degrees Celsius (3.6 degrees Fahrenheit). At the same time, the isotope ratio of carbon in the water and atmosphere shifted to correspond to the values observed in the fossils. Moreover, this carbon isotope ratio gradually returned to normal within 200,000 years, just as it does in actual fossil records.

Dickens's model is compelling but rare. So far the significance of methane from natural gas hydrate sources as a greenhouse gas has received only limited consideration in global climate modeling. The contribution of methane hydrates to global carbon budgets has likewise not been adequately taken into account.

Trying to come up with data to correct these shortfalls is one of the greatest motivations for our work on methane hydrates at GEOMAR. We continue to focus on Hydrate Ridge off the Oregon coast—eight marine expeditions targeted the site earlier this year. We will also be looking at the seafloor off the coasts of Costa Rica, Nicaragua, Alaska's Aleutian Islands and New Zealand. Whatever happens in methane hydrate research, the future will be anything but dull.

The Authors

Erwin Suess, Gerhard Bohrmann, and Jens Greinert work together in the department of marine environmental geology at the Research Center for Marine Geosciences in Kiel, Germany. Suess has managed the department since its inception in 1988 and has directed the research center since 1995. Bohrmann is director of the center's core repository, an extensive collection of marine sediment samples available for scientific analysis. Greinert is a postdoctoral researcher. Together they make up one of the world's leading groups studying methane hydrates and submarine cold springs. Science writer Erwin Lausch is a member of the research center's advisory counsel.

Technology for collecting solar energy has advanced greatly since photovoltaic cells, which transform the sun's rays into electricity, were invented in the 1950s. The cells' initially high cost meant that for decades they were not widely used except on satellites and in rural areas where stringing a power line was prohibitively expensive. Today, however, one often sees solar panels on the roofs of houses and commercial buildings, and solar power plants operate in the United States and several other nations.

In the broadest definition, however, solar energy encompasses a range of technologies and natural processes far beyond the mechanical capture of sunlight and its conversion to power in photovoltaic cells. For example, heat engines capture the warmth of the sun, which is then converted to electricity. Through photosynthesis, sunlight provides the energy plants need to grow; burning agricultural waste or other plant matter releases this energy. The uneven heating of the earth's surface by the sun gives rise to wind, which can be harnessed by wind turbines. The following article looks at these varied methods of using the sun to meet the earth's energy needs. —LEH

"Solar Energy"
by William Hoagland
Scientific American, September 1995

Every year the earth's surface receives about 10 times as much energy from sunlight as is contained in all the known reserves of coal, oil, natural gas and uranium combined. This energy equals 15,000 times the world's annual consumption by humans. People have been burning wood and other forms of biomass for thousands of years, and that is one way of tapping solar energy. But the sun also provides hydropower, wind power and fossil fuels—in fact, all forms of energy other than nuclear, geothermal and tidal.

Attempts to collect the direct energy of the sun are not new. In 1861 a mathematics instructor named Augustin-Bernard Mouchot of the Lycée de Tours in France obtained the first patent for a solar-powered motor. Other pioneers also investigated using the sun's energy, but the convenience of coal and oil was overwhelming. As a result, solar power was mostly forgotten until the energy crisis of the 1970s threatened many major economies.

Economic growth depends on energy use. By 2025 the worldwide demand for fuel is projected to increase by 30 percent and that for electricity by 265 percent. Even with more efficient use and conservation, new sources of energy will be required. Solar energy could provide 60 percent of the electricity and as much as 40 percent of the fuel.

Extensive use of more sophisticated solar energy technology will have a beneficial impact on air pollution and global climatic change. In developing countries, it can alleviate the environmental damage caused by the often inefficient practice of burning plant material for cooking and heating. Advanced solar technologies have the potential to use less land than does biomass cultivation: photosynthesis typically captures less than 1 percent of the available sunlight, but modern solar technologies can, at least in the laboratory, achieve efficiencies of 20 to 30 percent. With such efficiencies, the U.S. could meet its current demand for energy by devoting less than 2 percent of its land area to energy collection.

It is unlikely that a single solar technology will predominate. Regional variations in economics and the availability of sunlight will naturally favor some approaches over others. Electricity may be generated by burning biomass, erecting wind turbines, building solar-powered heat engines, laying out photovoltaic cells or harnessing the energy in rivers with dams. Hydrogen fuel can be produced by electrochemical cells or biological processes—involving microorganisms or enzymes—that are driven by sunlight. Fuels such as ethanol and methanol may be generated from biomass or other solar technologies.

Solar energy also exists in the oceans as waves and gradients of temperature and salinity, and they, too, are potential reservoirs to tap. Unfortunately, although the energy stored is enormous, it is diffuse and expensive to extract.

Growing Energy

Agricultural or industrial wastes such as wood chips can be burned to generate steam for turbines. Such facilities are competitive with conventional electricity production wherever biomass is cheap. Many such plants already exist, and more are being commissioned. Recently in Värnamo, Sweden, a modern power plant using gasified wood to fuel a jet engine was completed. The facility converts 80 percent of the energy in the wood to provide six megawatts of power and nine megawatts of heat for the town. Although biomass combustion can be polluting, such technology makes it extremely clean.

Progress in combustion engineering and biotechnology has also made it economical to convert plant material into liquid or gaseous fuels. Forest products, "energy crops," agricultural residues and other wastes can be gasified and used to synthesize methanol. Ethanol is released when sugars, derived from sugarcane or various kinds of grain crops or from wood (by converting cellulose), are fermented.

Alcohols are now being blended with gasoline to enhance the efficiency of combustion in car engines and to reduce harmful tail-pipe emissions. But ethanol can be an effective fuel in its own right, as researchers in Brazil have demonstrated. It may be cost-competitive with gasoline by 2000. In the future, biomass plantations could allow such energy to be "grown" on degraded land in developing nations. Energy crops could also allow for better land management and higher profits. But much research is needed to achieve consistently high crop yields in diverse climates.

Questions do remain as to how useful biomass can be, even with technological innovations. Photosynthesis is inherently inefficient and requires large supplies of water. A 1992 study commissioned by the United Nations concluded that 55 percent of the world's energy needs could be met by biomass by 2050. But the reality will hinge on what other options are available.

Wind Power

Roughly 0.25 percent of the sun's energy reaching the lower atmosphere is transformed into wind—a minuscule part of the total but still a significant source of energy. By one estimate, 80 percent of the electrical consumption in the U.S. could be met by the wind energy of North and South Dakota alone. The early problems surrounding the reliability of "wind farms" have now been by and large resolved, and in certain locations the electricity produced is already cost-competitive with conventional generation.

In areas of strong wind—an average of more than 7.5 meters per second—electricity from wind farms costs as little as $0.04 per kilowatt-hour. The cost should drop to below $0.03 per kilowatt-hour by the year 2000. In California and Denmark more than 17,000 wind turbines have been completely integrated into the utility grid. Wind now supplies about 1 percent of California's electricity.

One reason for the reduction is that stronger and lighter materials for the blades have allowed wind machines to become substantially larger. The turbines now provide as much as 0.5 megawatt apiece. Advances in variable-speed turbines have reduced stress and fatigue in the moving parts, thus improving reliability.

5.0

Over the next 20 years better materials for air foils and transmissions and smoother controls and electronics for handling high levels of electrical power should become available.

One early use of wind energy will most likely be for islands or other areas that are far from an electrical grid. Many such communities currently import diesel for generating power, and some are actively seeking alternatives. By the middle of the next century, wind power could meet 10 to 20 percent of the world's demand for electrical energy.

The major limitation of wind energy is that it is intermittent. If wind power constitutes more than 25 to 45 percent of the total power supply, any shortfall causes severe economic penalties. Better means of energy storage would allow the percentage of wind power used in the grid to increase substantially. (I will return to this question presently.)

Heat Engines

One way of generating electricity is to drive an engine with the sun's radiant heat and light. Such solar-thermal electric devices have four basic components, namely, a system for collecting sunlight, a receiver for absorbing it, a thermal storage device and a converter for changing the heat to electricity. The collectors come in three basic configurations: a parabolic dish that focuses light to a point, a parabolic trough that focuses light to a line and an array of flat mirrors spread over several acres that reflect light onto a single central tower.

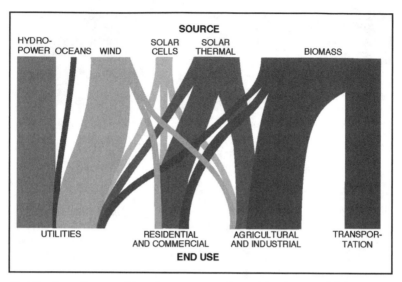

Distribution of renewable solar energy projected for the year 2000 shows that many different means of tapping the resource will play a role.

These devices convert between 10 and 30 percent of the direct sunlight to electricity. But uncertainties remain regarding their life span and reliability. A particular technical challenge is to develop a Stirling engine that performs well at low cost. (A Stirling engine is one in which heat is added continuously from the outside to a gas contained in a closed system.)

Solar ponds, another solar-thermal source, contain highly saline water near their bottom. Typically, hot water rises to the surface, where it cools off. But salinity makes the water dense, so that hot water can stay at the bottom and thus retain its heat. The pond traps the sun's radiant heat, creating a high temperature gradient. Hot, salty fluid is drawn out from the bottom of the pond and allowed to evaporate; the vapor is used to drive a Rankine-cycle engine similar to that installed in cars. The cool liquid at the top of the pond can also be used, for air-conditioning.

A by-product of this process is freshwater from the steam. Solar ponds are limited by the large amounts of water they need and are more suited to remote communities that require freshwater as well as energy. Use of solar ponds has been widely investigated in countries with hot, dry climates, such as in Israel.

Solar Cells

The conversion of light directly to electricity, by the photovoltaic effect, was first observed by the French physicist Edmond Becquerel in 1839. When photons shine on a photovoltaic device, commonly made of silicon, they eject electrons from their stable positions, allowing them to move freely through the material. A voltage can then be generated using a semiconductor junction. A method of producing extremely pure crystalline silicon for photovoltaic cells with high voltages and efficiencies was developed in the 1940s. It proved to be a tremendous boost for the industry. In 1958 photovoltaics were first used by the American space program to power the radio of the *U.S. Vanguard I* space satellite with less than one watt of electricity.

Although significant advances have been made in the past 20 years—the current record for photovoltaic efficiency is more than 30 percent—cost remains a barrier to widespread use. There are two approaches to reducing the high price: producing cheap materials for so-called flat-plate systems, and using lenses or reflectors to concentrate sunlight onto smaller areas of (expensive) solar cells. Concentrating systems must track the sun and do not use the diffuse light caused by cloud cover as efficiently as flat-plate systems. They do, however, capture more light early and late in the day.

Virtually all photovoltaic devices operating today are flat-plate systems. Some rotate to track the sun, but most have no moving parts. One may be optimistic about the future of these devices because commercially available efficiencies are well below theoretical limits and because modern manufacturing techniques are only now being applied. Photovoltaic electricity produced by either means should soon cost less than $0.10 cents per kilowatt-hour, becoming competitive with conventional generation early in the next century.

Storing Energy

Sunlight, wind and hydropower all vary intermittently, seasonally and even daily. Demand for energy fluctuates as well; matching supply and demand can be accomplished only with storage. A study by the Department of Energy estimated that by 2030 in the U.S., the availability of appropriate storage could enhance the contribution of renewable energy by about 18 quadrillion British thermal units per year.

With the exception of biomass, the more promising long-term solar systems are designed to produce only electricity. Electricity is the energy carrier of choice for most stationary applications, such as heating, cooling, lighting and machinery. But it is not easily stored in suitable quantities. For use in transportation, light-weight, high-capacity energy storage is needed.

Sunlight can also be used to produce hydrogen fuel. The technologies required to do so directly (without generating electricity first) are in the very early stages of development but in the long term may prove the best. Sunlight falling on an electrode can produce an electric current to split water into hydrogen and oxygen, by a process called photoelectrolysis. The term "photobiology" is used to describe a whole class of biological systems that produce hydrogen. Even longer-term research may lead to photocatalysts that allow sunlight to split water directly into its component substances.

When the resulting hydrogen is burned as a fuel or is used to produce electricity in a fuel cell, the only by-product is water. Apart from being environmentally benign, hydrogen provides a way to alleviate the problem of storing solar energy. It can be held efficiently for as long as required. Over distances of more than 1,000 kilometers, it costs less to transport hydrogen than to transmit electricity. Residents of the Aleutian Islands have developed plans to make electricity from wind turbines, converting it to hydrogen for storage. In addition, improvements in fuel cells have allowed a number of highly efficient, nonpolluting uses of hydrogen to be developed, such as electric vehicles powered by hydrogen.

A radical shift in our energy economy will require alterations in the infrastructure. When the decision to change is made will depend on the importance placed on the environment, energy security or other considerations. In the U.S., federal programs for research into renewable energy have been on a roller-coaster ride. Even the fate of the Department of Energy is uncertain.

At present, developed nations consume at least 10 times the energy per person than is used in developing countries. But the demand for energy is rising fast everywhere. Solar technologies could enable the developing world to skip a generation of infrastructure and move directly to a source of energy that does not contribute to global warming or otherwise degrade the environment. Developed countries could also benefit by exporting these technologies—if additional incentives are at all necessary for investing in the future of energy from the sun.

The Author

William Hoagland received an M.S. degree in chemical engineering from the Massachusetts Institute of Technology. After working for Syntex, Inc., and the Procter & Gamble Company, he joined the National Renewable Energy Laboratory (formerly the Solar Energy Research Institute) in Golden, Colo., where he managed programs in solar materials, alcohol fuels, biofuels and hydrogen. Hoagland is currently president of W. Hoagland &

Associates, Inc., in Boulder. The editors would like to acknowledge the assistance of Allan Hoffman of the Department of Energy.

A photovoltaic cell collects sunlight—which is radiation in the visible part of the electromagnetic spectrum—and converts it directly to electricity. A thermophotovoltaic cell is similar, except it converts infrared radiation—better known as heat—into electricity. Many common appliances, such as ovens and water heaters, transform electricity into heat. Thermophotovoltaics reverse this process, providing power from heat.

Burning some sort of fuel generally provides the heat required for thermophotovoltaic generators. Most thermophotovoltaics have been developed to use fossil fuels, but researchers are also looking at the suitability of burning biomass (plant materials and animal waste) or using highly concentrated solar rays as potential renewable fuels.

The following article discusses the potential of thermophotovoltaics to pick up where standard photovoltaics leave off. Since it was published in 1998, academic researchers have continued to develop thermophotovoltaic generators to

recharge the batteries in hybrid gas-electric
vehicles. The Midnight Sun thermophotovoltaic
generator, made by JX Crystals for use on sail-
boats, never progressed beyond the prototype
stage, however. More funding and more research
will be necessary to make thermophotovoltaics a
significant player in the energy market. —LEH

"Thermophotovoltaics"
by Timothy J. Coutts and Mark C. Fitzgerald
Scientific American, September 1998

Photovoltaics is a technology that typically transforms
sunlight into electricity. Radiation from the visible part
of the spectrum is, after all, abundant, nonpolluting and
free. But photovoltaics can also provide useful amounts
of electricity from infrared radiation—that is, radiant
heat generated by a source of energy such as fuel oil.

This lesser-known approach, called thermophotovoltaics, offers a major advantage in certain settings: a
generator can operate at night or when the sky is over-
cast, thereby eliminating any need for batteries to store
electricity. The technology is also preferable in some
ways to conventional electricity-generating technology
based on burning fossil fuels. Its efficiency—the percent
of fuel energy converted to electricity—can be substantially higher than that of electric generators powered
by natural gas or another fossil fuel. Moreover, a semi-
conductor-based thermophotovoltaic system can be
designed to minimize pollutants. And because it contains

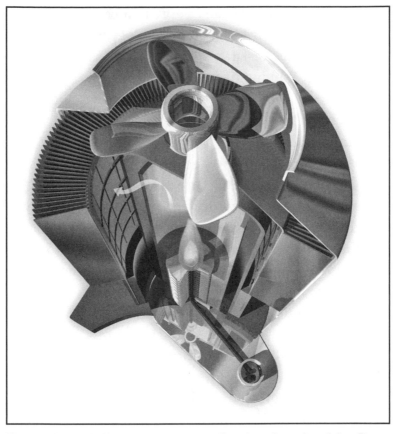

Thermophotovoltaic generator turns radiant heat into electricity. The burner (*center*) generates heat, whose infrared photons hit a radiator (*enclosure*). The radiator channels certain infrared wavelengths to arrays of photovoltaic cells (*grid*) that convert the energy to electricity.

no moving parts, it will run silently and reliably, requiring little maintenance.

In spite of these advantages, thermophotovoltaic technology has not enjoyed the success of solar photovoltaics, which today constitutes a thriving, albeit specialized, segment of the energy market. But the divide may soon shrink. The technology, which has its roots in the same research that produced solar cells, appears to be coming into its own.

The ideas that underlie thermophotovoltaics stretch back 40 years. Pierre R. Aigrain of the École Normale Supérieure in Paris first described some of the basic concepts during a series of lectures in 1956. In the early 1960s U.S. Army researchers at Fort Monmouth, N.J., created the first documented prototype of a thermophotovoltaic generator. Its efficiency was less than 1 percent, however. Efficiencies of 10 to 15 percent were needed for a usable generator that could be deployed by troops operating in the field.

Throughout the late 1970s and into the early 1980s, research supported and carried out by the Electric Power Research Institute in Palo Alto, Calif., the Gas Research Institute in Chicago, Stanford University and other organizations achieved some improvement in performance. But the components of early systems could never channel enough heat to the units that convert infrared energy into electricity. Recently a new set of materials has taken the technology past the development stage.

Commercial Debut

Thermophotovoltaics is about to reach the commercial marketplace. A company in the Pacific Northwest plans to market a thermophotovoltaic generator to run electrical equipment on sailboats. Other applications under development include small power units that would supply electricity in remote areas or for roving military troops. The technology could also assist in running hybrid electric vehicles, in which an electric battery complements the power from an internal-combustion engine. Ultimately, thermophotovoltaics could produce megawatts of power, furnishing some of the requirements of utilities or helping industry meet its electricity needs by recovering the unused heat from industrial processes.

Production of electricity from radiant heat requires several functional elements. A source of heat must be coupled with a radiator, a material that emits desired wavelengths of infrared radiation. A semiconductor device consisting of an array of interconnected cells must be engineered to convert these select wavelengths to electricity, which is then sent through a circuit to perform useful work—running a refrigerator on a boat, for instance. Finally, to perform efficiently, a thermophotovoltaic system must rechannel unused energy back to the radiator. In some applications the waste energy can also be used for other purposes, such as heating a room.

Sources of heat for a thermophotovoltaic system might range from fossil-fuel burners to sunlight to a nuclear fission reaction. As a practical matter, most systems under development deploy fossil fuels. Solar energy boosted to high intensities by "concentrator" devices can be used to operate a thermophotovoltaic generator; but the room-size concentrators and devices needed to store heat for nighttime use are still in the early development stages. Many experts consider the other alternative, nuclear fuel, less acceptable because of public fears of radioactivity. Thus, the nuclear option may be relegated to highly specialized applications, such as unmanned space probes to the outer reaches of the solar system, where solar cells would cease to function for lack of sunlight.

In constructing a burner to supply heat, researchers have begun to consider taking advantage of the metal-mesh or ceramic containers used in industrial drying of paper, inks, paints and agricultural crops. With a large surface area, these burners achieve the necessary temperatures of above 1,000 degrees Celsius (1,832 degrees Fahrenheit).

Radiators are needed because the semiconductor converter that transforms radiant heat to electricity cannot efficiently use the infrared energy produced by combusted fuel. The converters can operate efficiently only within a specific range of wavelengths, whereas the heat from a flame transmits infrared energy at wavelengths and intensities that can shift unpredictably (because the flame is subject to air currents and varia-tions in temperature). The radiators transform the available heat energy into a well-defined range of wave-lengths of uniform intensity.

A radiator may be built as a flat or rounded surface or as an array of tiny filaments. The material properties of oxides of rare-earth elements such as ytterbium, erbium and holmium allow heat to be radiated from the burner in relatively narrow bands of wavelengths; silicon carbide, in contrast, emits a broader spectrum.

In constructing a converter, thermophotovoltaic researchers choose semiconductors that match the infrared spectrum of the wavelengths emitted by a radiator. These wavelengths correspond roughly to the energy needed to free electrons that would otherwise remain bound within the crystalline solid of the semiconductor converter [*see box on page 180*]. The most energetic immobile electrons reside in the so-called valence band of the semiconductor crystal, which describes the range of allowed energy levels of the outermost bound electrons. Electrons in the valence band are not free to move through the crystal. When an atom absorbs an infrared photon of just the right amount of energy, an electron is boosted to the conduction band, where it can flow in a current in the crystal. (A photon is a unit, or quantum, of electromagnetic energy. According to the laws of quantum mechanics, all electromagnetic energy has the properties of both a wave and a particle, and the energy of a photon determines the wavelength of its corresponding wave.) The photon energy required to move an electron from the valence to the conduction band is known as the band-gap energy and is expressed in electron volts.

Once in the conduction band, electrons cross a junction between two dissimilar areas of the semiconductor crystal, moving more freely in one direction than the other. The resulting congregation of negatively charged electrons on one side of the junction causes a

Converter cell transforms heat into electricity when infrared photons with the right energy penetrate the cell near the intersection of two unlike areas in a semiconductor crystal. When a photon encounters an atom there, it dislodges an electron and leaves a hole. The electron migrates into the n-type layer (one that has more electrons than holes), and the holes move into the p-type layer (one that has more holes than electrons). The electron then moves into an electrical contact on the cell and travels through an external circuit until it reappears in the p-type layer, where it recombines with a hole. If the photon has less than the desired energy, or band gap, it is reflected off the quartz shield and back to the radiator.

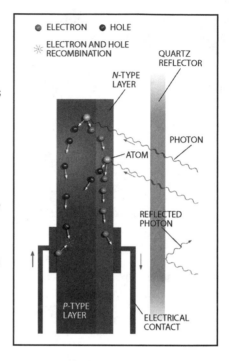

buildup of negative electrical potential, which acts as a force pushing the electrons in a current that flows through the photovoltaic cell. Each converter consists of a series of cells wired together to increase the power output. The current generated from the converter can then move through a wire connected to a lamp or a household electrical system.

Advances in converter materials have enabled engineers to improve electrical output. They can select converters that because of their particular band gaps respond best to the range of wavelengths emitted by a given radiator. In past years, the absence of appropriate radiator-converter combinations had posed a key obstacle to furthering the technology.

The first generation of thermophotovoltaic devices used radiators that produced a narrow band of wavelengths. Radiators made of the rare-earth material ytterbium oxide often accompanied silicon semiconductor converters with a band gap of 1.14 eV. In theory, a selective radiator—which emits a narrow band of wavelengths—should be more efficient than broader-spectrum (broadband) designs. The photons in a selective radiator should provide the minimum energy needed to lift an electron in the semiconductor converter into the conduction band; any excess energy would otherwise be lost as waste heat. Selective radiators should thus supply more power at a lower cost per watt of electricity. In practice, the systems have never performed as expected. The radiators fail to emit enough of the energy from the burning fuel at the precise wavelength needed by a material like silicon to make the conversion process efficient or to produce enough power.

Moreover, temperatures of 2,000 degrees C are required to provide sufficient intensity to achieve worthwhile power output. The heat can stress the material composing the radiator as well as other components, thereby shortening its life span. In addition, polluting nitrous oxide emissions can also result from fuel combustion at these torrid temperatures.

Thermophotovoltaics has advanced because researchers have learned how to pair radiators that transmit a relatively broad range of wavelengths with semiconductors able to cope with that wide spectrum.

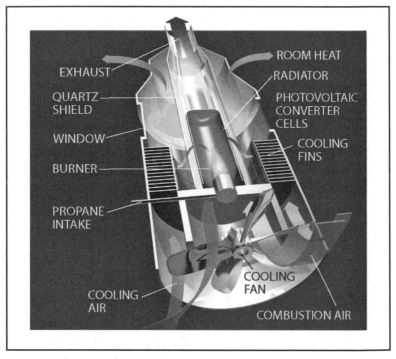

Generator (*diagram*) consists of a burner that combines fuel and air to produce heat and of an infrared radiator surrounded by thermophoto-voltaic converter cells with attached cooling fins. When the fuel ignites, the radiator heats up to 1,250 degrees Celsius or more. Then 48 gallium antimonide cells, connected in series, convert the infrared energy emitted by the radiator to electric power. At the same time, some of the current starts the fan, which blows air upward to cool the photovoltaic cells. Excess electric power produced by the generator is delivered to a battery for later use. At its optimal operating temperature, this circuit can supply 30 watts of power. Extra heat can be used to warm a room.

Broadband radiators, such as silicon carbide, are capable of working efficiently at cooler temperatures of up to 1,000 degrees C. Semiconductor materials developed for the solar energy industry from the third and fifth columns of the periodic table—so-called III-V materials, such as gallium antimonide and indium gallium arsenide—perform the photovoltaic conversion at the wavelengths emitted by these radiators. The band-gap energy required to generate power from III-V materials—0.5 to 0.7 eV—is much less than the 1.14 eV necessary for silicon.

No thermophotovoltaic system can convert all the infrared energy to electricity. Any photon with energy lower than the band gap of the converter cannot boost an electron from the valence to the conduction band and so does not generate electricity. These unused photons become waste heat, unless some means can be found to use them. A photon-recuperation system is a component of a thermophotovoltaic system that sends lower-energy photons back to the radiator, which reabsorbs them, helping to keep the radiator heated and to conserve fuel. Then, more of the emitted photons reach or exceed the band gap.

Retrieving Unused Photons

Investigators have explored several approaches to photon recovery, including the use of a grid of microscopic metal antennae. These antennae, which might be a thin metal film atop a converter cell, transmit the desired infrared wavelengths to the converter while reflecting other photons back to the radiator. Many photon-recovery schemes have fallen short: some systems detect too narrow a range of wavelengths; others are too expensive. The most promising option is the back-surface reflector—so named because the unabsorbed photons move entirely through the layers of the semi-conductor converter and are then returned to the radiator by a highly reflective gold surface on the back of the converter.

Around the world, researchers are exploring various technical avenues for developing and commercializing thermophotovoltaics. Since 1994 they have gathered for three international conferences sponsored by the U.S. Department of Energy's National Renewable Energy Laboratory (NREL). The Defense Advanced Research Projects Agency (DARPA), the DOE and the U.S. Army Research Office all have funded programs to develop the technology.

Scientists are excited by laboratory modeling that suggests the promise of this technology. With a radiator that can operate at 1,500 degrees C, it appears possible for semiconductor cells with a single junction (the site where electric potential builds up) to obtain power-density outputs of three to four watts per square centimeter of converter area.

Cells with multiple junctions are also being considered, an approach borrowed from the solar photovoltaics industry. Multijunction cells would allow the converter to capture a wider spectrum of wavelengths, thereby making better use of broadband radiators. Each of the separate junctions would generate a current after absorbing photons in different energy ranges. Theoretically, multijunction devices might achieve five to six watts of energy per square centimeter. Compare these projections with those for the typical flat-panel solar-cell array, which typically achieves 15 milliwatts per square centimeter. Whereas these estimates derive from computer models, and actual power outputs will undoubtedly be smaller, prototypes have exhibited power densities greater than one watt per square centimeter.

The slow and painstaking work of designing and integrating the component parts of a thermophotovoltaic system has begun at several private and government laboratories. To obtain usable amounts of power, each converter cell must be wired to other cells. Traditional semiconductor manufacturing methods can pattern, etch and wire together multiple cells on a single surface, called a wafer. Researchers at the NREL and those working separately at Spire Corporation in Bedford, Mass., and the National Aeronautics and Space Administration Lewis Research Center have demonstrated these techniques for arrays of thermo-photovoltaic cells. One notable example comes from NREL researchers Scott Ward and Mark Wanlass, who have interconnected small thermophotovoltaic cells, each wired in series. The wiring connections move along the top of one cell and then run down along the back of an adjacent one in a configuration that reduces current, increases voltage and minimizes power loss.

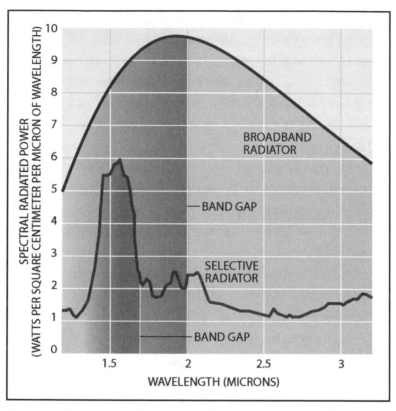

Photon radiators emit either a narrow band of wavelengths (*selective*) or a wider range (*broadband*) that can be transformed into electricity by a semiconductor converter. Selective radiators, made of rare-earth materials, produce a smaller number of photons at or above the band gaps at which electricity can be generated (*dark gray area*). Broadband radiators, in contrast, can take advantage of more of the usable power reaching the cell (*medium gray area*). Lower-energy photons—those to the right of the band gap—become waste heat.

This design could eventually reduce the converter to a wafer populated with cells that have only two external contacts—the ones needed to create a circuit—that could power a water pump or a cabin in the

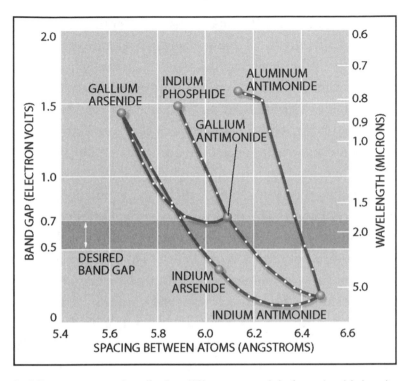

Building converters by alloying different materials from the third and fifth columns of the periodic table can achieve a desired band gap—the amount of energy needed to free an electron so it can flow in a current. Band gaps of 0.5 to 0.7 eV (*shaded area*) are ideal for thermo-photovoltaic devices and correspond to increasingly shorter infrared wavelengths from the radiator. These band gaps can be obtained by combining compounds such as gallium antimonide with indium anti-monide in increments represented by the dots along the dark gray lines. Calculating the spacing between atoms (*bottom axis*) allows designers to join together different compounds with distinct crystalline structures.

woods. Many such wafers could be connected to achieve desired power requirements. The integration of photovoltaic cells may reduce the high cost of the technology, because the cells can be manufactured using standard semiconductor manufacturing methods. In their prototype, Ward and Wanlass also devised a novel design for recirculating unused photons. The electrically active areas of the wafer rest atop an indium phosphide substrate that is semi-insulating. Because the material is nonconductive, and most electrons remain relatively tightly bound within the semiconductor crystal, low-energy photons can move through the substrate without getting absorbed by free electrons moving through the conduction band. The photons then reflect off a gold surface and return to the radiator. In other prototype designs, many unused photons are absorbed by the converters.

While developmental work continues elsewhere, the first thermophotovoltaic commercial product is about to reach the market. JX Crystals in Issaquah, Wash., has created a product—Midnight Sun—primarily for use on sailboats. The 14-centimeter-wide-by-43-centimeter-tall cylindrical heater, powered by propane gas, can produce 30 watts of electricity and is targeted as a means of recharging batteries that run navigation and other equipment. The unit not only provides electricity but acts as a co-generator, supplying space heating for the boat cabin. It uses a partially selective radiator made of magnesium aluminate and has gallium antimonide photovoltaic cells connected in series.

Although its current $3,000 price tag makes it more expensive than a conventional diesel generator, Midnight Sun runs silently and is expected to be more reliable, because it lacks any moving parts.

The product may also prove attractive to owners of recreational vehicles or wilderness homes, who could take advantage of a substantially less costly unit than the stainless-steel and brass thermophotovoltaic generator necessary for the marine environment.

Despite the present drawbacks of selective radiator systems, researchers still study them. DARPA has funded Thermo Power's Tecogen division in Waltham, Mass., to create a gas-fired generator as a power supply for troop communications or for powering laptop computers in the field. In the civilian realm the same unit could keep a home furnace running in a power outage. The 150- and 300-watt-power modules use arrays of ytterbium oxide fibers at a wavelength of 980 nanometers. These selective radiators are matched with silicon photoconverters. A multiple-layer insulating filter recovers unused energy.

Beyond the Niche

Though still in its infancy—actually just barely out of the research laboratory—thermophotovoltaics holds great promise for many niche markets. Longer term, the technology could play a broader role in the global energy market. The terminology "niche market" can itself be deceiving, evoking the image of a few buyers served by a small cottage industry. Yet the world is replete with billion-dollar niche markets—many of which had humble beginnings. Thomas J. Watson, the founder of IBM, originally saw a market for only a small handful of computers worldwide but decided to pursue the opportunity anyway.

The recovery of industrial waste heat could establish a huge market for thermophotovoltaics. Many industrial sectors—manufacturers of glass, aluminum, steel and other products—generate enormous quantities of heat through their production processes. The glass industry estimates that two thirds of the energy it consumes emanates as waste heat, which could amount to a gigawatt of power. Thermophotovoltaic converters could possibly be used to generate electricity from waste heat, affording enormous savings in electricity costs. Another promising application is suggested by a hybrid electric vehicle introduced by Western Washington University. The vehicle supplements the power from electric batteries by supplying 10 kilowatts from a thermophotovoltaic generator.

Today the funding devoted to thermophotovoltaics is not more than $20 million to $40 million a year. That is the money researchers receive for their studies on converter devices, radiators and the integration of the various components into working generators. A recent market study financed by the NREL, however, indicates that thermophotovoltaics could produce $500 million in commercial sales by 2005. Revenues for this nascent market would come from substituting this technology for diesel generators of less than two kilowatts used in the military and in recreational vehicle and boating markets. Thermophotovoltaics may lead to the possibility of cleaner, more efficient and inexpensive solutions for a number of alternative energy markets. A technology that had its roots in the 1950s may finally get a chance to prove itself at the turn of the new millennium.

The Authors

Timothy J. Coutts and Mark C. Fitzgerald have worked on renewable energy and related technologies for many years. Coutts, who has studied solar-cell technology since 1971, is a research fellow in the National Center for Photovoltaics of the U.S. Department of Energy's National Renewable Energy Laboratory (NREL) in Golden, Colo. He was educated at Newcastle Polytechnic in Britain, where in 1968 he received his doctorate in thin-film physics. At the NREL, he has broad-ranging responsibilities in both solar and thermophotovoltaics technologies. Fitzgerald is president of Science Communications, a photovoltaic consulting firm in Highlands Ranch, Colo. He has 19 years of experience providing education services about renewable energy.

The following article discusses several utilities that offer their customers green energy—power derived from renewable sources rather than traditional fossil fuel and nuclear plants. Since the article was published in 1997, such programs have taken off. More than 20,000 customers are part of the Sacramento Municipal Utility District's Greenergy program. Colorado Public Service Company was bought by Xcel Energy, whose Windsource program provides renewable energy to 36,000 customers in three states. In 2003, more than 10,500 customers took part in the Energy for Tomorrow renewable energy program of Wisconsin Electric (now known as We Energies).

Overall, about 50 percent of retail customers in the United States have the option to purchase green power through their electricity supplier, according to the U.S. Department of Energy. These sales have been increasing about 30 percent each year, according to the National Renewable Energy Laboratory.

In states with electricity competition, customers can purchase "green power" from an alternative electricity supplier. More than 500 regulated utilities in other states offer power from green sources as well as traditional ones. Green power is almost always more expensive than that from traditional sources, but environmentally aware Americans are quite willing to pay the extra premium. —LEH

"Change in the Wind"
by W. Wayt Gibbs
Scientific American, October 1997

Most utilities offer as much choice in how your electricity is created as Henry Ford offered to those buying his Model T: you can have any color you want, as long as it is black. But as power companies face deregulation and the prospect of competing for customers, many are beginning to sell a second, distinctly greener stream of energy. The juice flowing from solar cells, windmills and biomass furnaces is still a mere trickle running into an ocean of fossil- and nuclear-fueled power. But pilot projects are revealing just how many people will pay more for electricity that pollutes less.

The tiny, city-owned utility that serves Traverse City, Mich., gambled that many of its customers would pay a 23 percent premium (typically about $7.50 a month) to light their lamps with wind rather than coal. With a grant from the state and a subsidy from the U.S. Department of Energy, the electric company erected a giant, 600-kilowatt windmill with blades 44 meters (144 feet) in diameter—the largest such turbine in North America.

Some 145 residents and 20 businesses signed up; another 75 filled a waiting list. "That amounts to 3 percent of our 8,000 customers," says Steve Smiley, who managed the project. Love of Mother Earth was not the only incentive for these people, he notes. "We also promised 'green' customers that we would not increase their rates in the future, since the fuel is free."

Several years ago the Sacramento Municipal Utility District began installing small photovoltaic panels on the roofs of those willing to pay an extra $4 a month. Thousands applied, but the panels cost about $20,000 apiece, so the company has so far set up only 420, enough to generate 1.7 megawatts. In May the utility signed contracts to add 10 megawatts' worth of solar cells over the next five years.

The company also kicked off a new green pricing program similar to Traverse City's: for an extra cent per kilowatt-hour, subscribers will get all their electricity from new renewable sources. (Not literally: green customers still draw power from every oil- and gas-fired dynamo on the grid. But their checks pay for cleaner generators.)

Some 23 other companies have followed suit. Public Service Company of Colorado has begun enlisting buyers for a 10-megawatt wind farm. Wisconsin Electric signed up more than 7,000 volunteers for hydroelectric and biomass power.

The trend is encouraging, says Blair G. Swezey of the National Renewable Energy Lab, but should not be mistaken for a resurgence in renewables. In fact, utilities are adding renewable capacity at just one fifth the rate they did a decade ago. Nonpolluting energy is closing in on the cost of coal and oil, but it is not there yet.

How close is close enough? In surveys, 40 to 60 percent say they would pay more for cleaner power. "But the story changes when people get their checkbooks out," observes Terry Peterson of the Electric Power Research Institute in Palo Alto, Calif. Few green-power programs have enrolled more than 5 percent of ratepayers. To be sure, most were poorly advertised and asked for premiums of 20 percent or more.

But an exception may prove to be the rule. When Massachusetts let homeowners in four cities choose among nine power vendors last summer, 16 percent chose Working Assets Green Power, which buys no electricity from nuclear or coal plants. Although Working Assets's rates were the highest of the nine competitors, they were still cheaper than the monopoly that customers were leaving. "For green pricing to make a real difference, you need to charge less than what people pay today," says Laura Scher, who managed the project.

That will be difficult, Swezey argues, as long as utilities can bill customers separately for failed investments, such as prematurely closed nuclear reactors. If those costs were instead factored into the price of electricity, then wind and dam power would look like more of a bargain. Because they are not, Swezey wagers it will take several years of healthy competition before the renewable power industry starts seeing green.

Web Sites

Due to the changing nature of Internet links, the Rosen Publishing Group, Inc., has developed an online list of Web sites related to the subject of this book. This site is updated regularly. Please use this link to access the list:

http://www.rosenlinks.com/saca/enpo

For Further Reading

Alternative Energy Institute. *Powering Our Future: An Energy Sourcebook for Sustainable Living.* Lincoln, NE: iUniverse Inc., 2005.

Berinstein, Paula. *Alternative Energy: Facts, Statistics, and Issues.* Westport, CT: Oryx Press, 2001.

Bodansky, David. *Nuclear Energy: Principles, Practices, and Prospects.* New York, NY: Springer, 2004.

Davis, Scott. *Microhydro: Clean Power from Water.* Gabriola Island, BC, Canada: New Society Publishers, 2003.

Deffeyes, Kenneth S. *Beyond Oil: The View from Hubbert's Peak.* New York, NY: Hill & Wang, 2006.

Ewing, Rex A. *Power with Nature: Solar and Wind Energy Demystified.* Masonville, CO: PixyJack Press, 2003.

Garwin, Richard L., and Georges Charpak. *Megawatts and Megatons: The Future of Nuclear Power and Nuclear Weapons.* Chicago, IL: University of Chicago Press, 2002.

Geller, Howard. *Energy Revolution: Policies for a Sustainable Future.* Washington, DC: Island Press, 2002.

Gipe, Paul. *Wind Energy Basics: A Guide to Small and Micro Wind Systems.* White River Junction, VT: Chelsea Green Publishing, 1999.

Gipe, Paul. *Wind Power: Renewable Energy for Home, Farm, and Business*. White River Junction, VT: Chelsea Green Publishing, 2004.

Goodell, Jeff. *Big Coal: The Dirty Secret Behind America's Energy Future*. New York, NY: Houghton Mifflin, 2006.

Heinberg, Richard. *Powerdown: Options and Actions for a Post-Carbon World*. Gabriola Island, BC, Canada: New Society Publishers, 2004.

Hoffmann, Peter. *Tomorrow's Energy: Hydrogen, Fuel Cells, and the Prospects for a Cleaner Planet*. Cambridge, MA: The MIT Press, 2002.

Morris, Robert C. *The Environmental Case for Nuclear Power: Economic, Medical, and Political Considerations*. New York, NY: Continuum International Publishing Group, 2000.

Murray, Raymond L. *Nuclear Energy: An Introduction to the Concepts, Systems, and Applications of Nuclear Processes*. Burlington, MA: Butterworth Heinemann, 2001.

Ramsey, Charles B., and Muhammad Modarres. *Commercial Nuclear Power: Assuring Safety for the Future*. Hoboken, NJ: John Wiley & Sons, 1998.

Ramsey, Dan. *The Complete Idiot's Guide to Solar Power for Your Home*. New York, NY: Alpha, 2002.

Roberts, Paul. *The End of Oil: On the Edge of a Perilous New World*. New York, NY: Mariner Books, 2005.

Schleer, Hermann. *The Solar Economy: Renewable Energy for a Sustainable Global Future*. London, England: Earthscan, 2002.

Sklar, Scott, and Kenneth Sheinkopf. *Consumer Guide to Solar Energy: New Ways to Lower Utility Costs, Cut Taxes, and Take Control of Your Energy Needs.* 3rd ed. Los Angeles, CA: Bonus Books, 2002.

Vaitheeswaran, Vijay V. *Power to the People: How the Coming Energy Revolution Will Transform an Industry, Change Our Lives, and Maybe Even Save the Planet.* New York, NY: Farrar, Straus and Giroux, 2005.

Bibliography

Anderson, Roger N. "Oil Production in the 21st Century." *Scientific American*, March 1998, pp. 86–91.

Ashley, Steven. "On the Road to Fuel-Cell Cars." *Scientific American*, March 2005, pp. 62–69.

Campbell, Colin J., and Jean H. Laherrère. "The End of Cheap Oil." *Scientific American*, March 1998, pp. 78–83.

Coutts, Timothy J., and Mark C. Fitzgerald. "Thermophotovoltaics." *Scientific American*, September 1998, pp. 90–95.

The Editors. "Disposing of Nuclear Waste." *Scientific American*, September 1995, p. 177.

Fouda, Safaa A. "Liquid Fuels from Natural Gas." *Scientific American*, March 1998, pp. 92–95.

Furth, Harold P. "Fusion." *Scientific American*, September 1995, pp. 174–176.

Gibbs, W. Wayt. "Change in the Wind." *Scientific American*, October 1997, p. 46.

Hoagland, William. "Solar Energy." *Scientific American*, September 1995, pp. 170–173.

Hollister, Charles D., and Steven Nadis. "Burial of Radioactive Waste Under the Seabed." *Scientific American*, January 1998, pp. 60–65.

Lake, James A., Ralph G. Bennett, and John F. Kotek. "Next-Generation Nuclear Power." *Scientific American*, January 2002, pp. 72–81.

Lloyd, Alan C. "The Power Plant in Your Basement." *Scientific American*, July 1999, pp. 80–86.

Schneider, David. "Power to the People." *Scientific American*, May 1997, p. 44.

Suess, Erwin, Gerhard Bohrmann, Jens Greinert, and Erwin Lausch. "Flammable Ice." *Scientific American*, November 1999, pp. 76–83.

Index

About the Editor

Linley Erin Hall is a writer and editor based in Berkeley, California. Her specialty is writing about science and engineering. She is also the editor of *The Laws of Motion: An Anthology of Current Thought* in the Contemporary Discourse in the Field of Physics series (Rosen, 2006). Hall has a graduate certificate in science communication from the University of California–Santa Cruz and a bachelor's degree in chemistry from Harvey Mudd College.

Illustration Credits

Cover Tim Boyle/Getty Images; p. 16 Jennifer Christiansen and Laurie Grace; pp. 20, 22–25 Laurie Grace (Source: Jean H. Laherrère); p. 27 Laurie Grace (Source: Petroconsultants, *Oil and Gas Journal* and U.S. Geological Survey); p. 29 Laurie Grace (Source: *Oil and Gas Journal*); pp. 33, 37, 39, 40, 41, 44 Daniels & Daniels; pp. 35, 39 (graph illustrations), 52, 55, 92–93, 100, 147 Laurie Grace; p. 38 (top) HITEC Drilling & Marine Systems; p. 38 (bottom) Laurie Grace (Source: Schlumberger); pp. 62, 66, 69, 71 Don Foley; pp. 83, 126 George Retseck; p. 89 William X. Haxby; pp. 104, 168 Carey Ballard; p. 131 Richard Hunt; pp. 151 (illustration), 175, 180, 182, 186, 187 Slim Films; p. 151 (photographs) courtesy of GEOMAR; p. 157 Rick Jones.

Series Designer: Tahara Anderson